Statistical Modeling for Naturalists

Statistical Modeling for Naturalists

By

Pedro F. Quintana Ascencio,
Federico López Borghesi
and Eric S. Menges

Cambridge
Scholars
Publishing

Statistical Modeling for Naturalists

By Pedro F. Quintana Ascencio, Federico López Borghesi
and Eric S. Menges

This book first published 2022

Cambridge Scholars Publishing

Lady Stephenson Library, Newcastle upon Tyne, NE6 2PA, UK

British Library Cataloguing in Publication Data
A catalogue record for this book is available from the British Library

ISBN (10): 1-5275-7587-X
ISBN (13): 978-1-5275-7587-5

CONTENTS

ACKNOWLEDGEMENTS

The main structure of this book took shape while teaching multiple iterations of statistical modeling classes to graduate students in the University of Central Florida, Orlando, Florida, USA, and short courses at El Colegio de La Frontera Sur, Chiapas, México, Universidad Técnica Particular de Loja, Loja, Ecuador, Universidad Regional Amazónica – Ikiam, Tena, Ecuador, and Universidad Rey Juan Carlos, Móstoles, Madrid, Spain. We are profoundly thankful to all participants for challenging our views. We thank James Angelo and Lina María Sánchez-Clavijo for their invaluable contributions. Reed Bowman kindly provided the front cover photo and Haoyu Li most of the black and white pictures. Conversations with Dave Jenkins, Carlos Iván Espinosa, Diego Vélez Mora, David Duncan, Betsey Boughton, Mario González Espinosa, Neptalí Ramírez Marcial, Brad Ochocki, Xavier Picó, Maria Paniw, Jesse Anaya, Aaron David, Will Crampton, Carl Weekley, Jennifer Navarra, Mark Burgman, Margaret Evans, John Fauth, Stephanie Koontz, David Nickerson, Fernando Quintana-Ascencio, José María Iriondo Alegría, Adrián Escudero, José Miguel Olano and Roberto Salguero Gómez were edifying. The support and advice of Eréndira Malinali Quintana Morales, Amarantha Zyanya Quintana Morales, Adrian Edward De Angelis, and Laura Cristina De Angelis have enriched this and many other ventures. Most especially, we are grateful for the unwavering patience and companionship of María Cristina Morales Hernández and Ericka Vanessa Correa Roldán without whom this book would not exist. We thank research assistants and interns at Archbold Biological Station and students at the University of Central Florida for their participation in collecting and organizing much of the data. The staff at Archbold Expeditions offered extensive aid and guidance. They are instrumental for the preservation of the ecosystems described in this book. We are particularly indebted to Luis Cayuela Delgado and Ian Biazzo for their extensive reviews and invaluable suggestions. They caught many errors and considerably improved the content of every chapter. Helen Edwards, our commissioning editor, provided continuous support during the final preparation of this document. We greatly celebrate the contributions of people that historically provided the conceptual basis allowing us to organize information, examine questions and guide our actions in efficient and creative ways. Any imprecisions and mistakes on the interpretation and implementation of the methods in this book are the responsibility and privilege of the authors.

FOREWORD

Statistical Modeling for Naturalists presents an outstanding introduction to different statistical methods focusing on the application of models for ecological problem-solving. It covers many statistical methods typically used in the field of ecology, such as linear models, generalized linear models and mixed modeling. The book provides access to datasets and code used for the different case studies. Two freely available, open-source software platforms are used in this book, R and Stan. R is one of the most popular programming languages for data analysis, statistical modeling, and visualization. Stan is a platform for statistical modeling and high-performance statistical computation. Both R and Stan run on all major platforms (Linux, Mac, Windows) and can interface with other data analysis languages.

But what makes this book unique compared with other published books featuring words such as "statistical" and "ecology"?

First, it is written by ecologists, and therefore the emphasis is on natural history. This is pretty well illustrated throughout the book, where each chapter develops a case study in population ecology based on decades of research conducted by the authors at Archbold Biological Station in Central Florida, USA. Each case study represents a research question, which is carefully framed by ecological theory about the problem at hand. This puts statistics at the service of ecological problem-solving and makes the book particularly appealing for biology and environmental science students and practitioners. By the end of the book, you will have acquired a pretty good knowledge of the ecological functioning of plant populations in the Florida scrub ecosystem even if, as in my case, you have never set foot in it.

Second, the book is based on a Bayesian approach to statistics; that is, statistical inference is based on the posterior distribution, which expresses all that is known about the parameters of a statistical model, given the data and existing knowledge. There are many advantages of the Bayesian approach, such as the possibility to combine previous with existing information to address the research question or the ease of error propagation, which is making the Bayesian paradigm gain tremendous momentum in statistics and its applications, including ecology. Yet, the

Bayesian approach might turn back many students who have been trained in the use of frequentist methods. But do not panic! The book provides very clear explanations about the way models are constructed and how to interpret the results. So regardless of you having received previous training either as a Bayesian or a frequentist, you will find the book very easy to follow. I must admit I am considering turning Bayesian myself after reading the book!

The way the book is structured makes it easy for readers to find good recipes for framing and analyzing their own research questions based on the presented case studies. But reading the book from beginning to end also gives an added value to the reader, as problems are presented in increasing complexity and quite often based on the results found in previous chapters. Statistical Modeling for Naturalists is worth reading. It offers insightful clues for conducting sound ecological research to students, practitioners, and researchers in different stages of their academic careers.

Enjoy the reading!

Luis Cayuela
Department of Biology and Geology, Physics, and Inorganic Chemistry
Universidad Rey Juan Carlos, Spain

Preface

This book aims to introduce the reader to captivating and useful methods for statistical data analysis and basic mathematical modeling. It is directed at people with rudimentary/intermediate statistical and mathematical instruction. It is neither a book for the beginner nor the specialist. For the former, we make our best efforts to explain every topic in the book in terms that are approachable to a novice. For those interested in mathematical demonstrations and programming specifics, we do our best to redirect them to more sophisticated sources. The book is written to appeal to naturalists like us. To this end, we focus on the questions and considerations that guide our modeling choices. We recapitulate three decades of ecological work on the Florida Scrub at Archbold Biological Station in Central Florida, USA. Each chapter follows two threads: one mathematical/statistical and the other ecological. This approach has been successful in retaining the attention of graduate students who are more interested in ecology and evolution than in data analysis and modeling. We are convinced about the prominence of mathematical and statistical thinking in producing robust, reproducible, and relevant science. But we are also persuaded of the need to constantly relate our models to the concepts being evaluated. The approaches in this book require code writing, so we use freely available software – R and Stan. We find these platforms to be among the best communitarian projects in recent decades, transforming the use and progress of mathematical and statistical applications. We maintain an associated website (https://github.com/StatsForNats/Book) where the reader can access all of the data and code used in the book. We must remark, however, that this book is not a guideline on either language. We recommend the reader to use the suggested literature to complement the material presented here. Most important, we employ Bayesian inference. Bayesian models are flexible and powerful but have been overlooked, mostly due to computational limitations and lack of training. Bayesian analyses formalize the process of evaluating information involved in the generation of knowledge, and they are particularly suited for the complex data used in ecological studies. With this book we intend to help make these tools accessible to naturalists and other people interested in understanding ecological mechanisms.

CHAPTER I

MODELS IN ECOLOGY

"All models are wrong, but some are useful"
– George Box

Relevant notions

Models allow us to explore complex phenomena and develop our understanding. They do this by offering a tradeoff – we must focus on a few aspects while excluding many others. Which factors we choose to include, and which to exclude, depends on what it is we are trying to understand. A model airplane might be useful to evaluate aerodynamic properties but it ignores the propulsion system.

The notion of models as simplified representations of complex reality is very much present across science. Such a vague definition of scientific models is purposeful. First, it allows us to recognize that models are ubiquitous to our way of thinking and can, therefore, take many forms. A scientific model can be anything from a visual model (such as a flowchart) to a computer program combining complex mathematical functions. In this book we deal mostly with statistical models, in which we propose a data-generating process and then estimate its parameters empirically. Regardless of its form, the purpose of any model is to make the phenomena in question easier to understand. The corollary of this statement is that we should not make models more complicated than they need to be. Second, this definition helps us to acknowledge the limitations of any given model. A frequent aphorism in statistics states that "all models are wrong, but some are useful." We prefer to see the glass half full and say that no model is fully correct. Models can provide different insights and help us to identify important information, but we must not feel guilty of letting go one model in favor of another. Finally, recognizing models as simplified descriptions of reality forces us to acknowledge that many assumptions are necessary in their construction. Model assumptions are the topic of Chapter V, but we want to emphasize here the importance of keeping them in mind and validating them whenever possible.

Within ecology, we make inaccurate observations on phenomena which arise from a myriad of interacting variables that change continuously. The size of a plant, for instance, is determined by multiple environmental factors (such as sunlight, water, and nutrients), biotic interactions (including diseases and predation), physical, chemical, and mechanical constraints, and the individual genetic background and ontogeny. The timing and intensity of each of these variables are affected by an increasing number of other factors.

Realistically, our studies can only deal with a small number of those factors, and models are the tools we use to help us focus on those relations. This is not to say that we randomly pick variables from a hat and use models to see whether they are helpful in explaining our observations. Our choices are biased and informed by our motivations, education, prior experiences, and ideas (Larson 2011). A butterfly flapping its wings might affect the initial conditions of a system sufficiently to help bring about a hurricane, but it would be silly for meteorologists to go around recording flying butterflies to forecast hurricanes. Their focus on data collected from weather tracking technology is well justified.

In this book we explore statistical approaches to evaluate ideas using data as arbitrator. When we propose models to formalize hypotheses (suggesting possible explanations of our observations), we require a way to substantiate them. One possible approach is through classical frequentist statistics. However, we agree with Howson and Urbach (1989) that the Newman-Pearson/Fisher approach fails to reach this goal in many realms. Their aspiration of objectivity, supposedly intended to avoid using degrees of belief, culminates in several paradoxes. These start with the problematic definition of "accepting" and "rejecting" a hypothesis since accepting it does not mean it is true and rejecting it does not imply it is false. The decision is arbitrary on the choice for the null hypothesis and its frequent lack of relevance for the question of interest. In any case, attempting to validate a hypothesis with the falsification of a null hypothesis does not inform how much the single alternative hypothesis compares to other logical explanations (Aho et al. 2014). Since they rely exclusively on the data at hand, ignoring any prior information, they are poorly suited for evaluating complex hypotheses and studies with limited data. There is an unwarranted belief in the certainty of a long-run frequency associated to the model parameters. Importantly, we believe the confidence placed in an arbitrary threshold as a single criterion when evaluating evidence is unjustified, and leads to poor inference (see Nuzzo 2014 and Wasserstein and Lazar 2016).

An alternative approach (the one embraced in this book) relies on the use of Bayesian inference. This approach ranks hypotheses within a spectrum of degrees of certainty in the light of evidence. It uses information available independently of a study alongside the likelihood of the data at hand to generate posterior probability distributions (see Chapter III). The posterior distributions from one study become available as prior information for future studies. Another advantage of Bayesian inference is that it considers model parameters as random variables instead of assuming they are true, fixed quantities (Ellison 2004). We find this long-standing philosophical tradition to be a satisfactory and robust way to evaluate ideas. It is a clear method for estimating model parameters and evaluating the degree of their uncertainty. It offers a probabilistic measure of the relative confidence we can have in competing models.

Touchon and McCoy (2016) documented that, while there has been a clear increase in the use of Bayesian statistics in the mainstream ecological literature, most doctoral programs in the United States still do not cover these approaches. In this book, we introduce Bayesian probabilistic frameworks as tools to create and evaluate models with the hope of reducing this disparity.

In his brilliant contribution, "Remedies for a scientific disease", Mark Burgman warns against a propensity among scientists to feel overly confident in our predictions. Since most ecological studies do not even attempt to determine if the intensity of monitoring is sufficient, there is a great risk of overlooking effects when the data are too limited. Often, many scientists wrongly interpret a lack of statistical effects as a lack of "real" effect. This is particularly damaging when assessing environmental impacts since the costs of false negatives can lead to real problems being disregarded. Burgman advocates for the use of the precautionary principle, weighting the costs of false positives and false negatives when evaluating the results of our analysis. We concord with him that one of the best safeguards against irrational interpretation is the use of images of the data and models' predictions in the form of scatterplots and histograms, and illustrations of estimates of associated uncertainty. With these ideas in mind, we try to present extensive images depicting the level of support for our models and encourage the reader to do the same.

Our model system is the Florida scrub and, in particular, an endangered and rare plant species inhabiting this ecosystem, both described in Chapter II. We use these model systems to introduce what we believe are powerful concepts that can help in developing descriptive statistics, inference, and modeling. We build these models as examples of tools to gain an understanding of ecological phenomena in this ecosystem. Using case studies dealing with contemporary ecological theories, we present relevant statistical concepts and procedures used to build models aimed to improve their understanding.

CHAPTER II

FLORIDA SCRUB

"The job of an ecologist is like putting together a giant puzzle, only all the pieces are hidden and the puzzle keeps changing"
– Warren Abrahamson

As natural ecosystems go, Florida scrub does not have a high profile. It is not majestic like a redwood forest, iconic as the Everglades, or as romantic as a midwestern grassland. But still, it is a beautiful, interesting, and at times confounding landscape. It is also the source of data used in this book. Here is a short description of Florida scrub.

Florida scrub is a shrubland, dominated by woody plants with multiple stems. It is not a forest (despite some prior categorizations of it as "sand pine scrub" or "sand pine forest"). The dominants are mainly shrub oaks, dwarf palmettos, and shrubs in the blueberry family. Elsewhere in the landscape, there are natural forests in Florida, which have the abundant precipitation necessary to support tree growth. But under most circumstances, Florida scrub has only scattered (or no) trees and is not routinely invaded by trees.

Florida scrub has sometimes been characterized as a desert, in part because there are many xeromorphic characteristics in its plants. These include small evergreen leaves that are thick and waxy, and stems that are often spiny. These traits are often assumed to be a function of low water availability but can also evolve in response to nutrient deficiency or herbivory (see Chapter III). Florida has no shortage of rain (although it is strongly seasonal) but the sandy soils supporting Florida scrub are extremely low in nitrogen, phosphorus, and other nutrients. With added nutrients, both planted citrus and introduced grasses can thrive on these soils. But, left undisturbed by humans, Florida scrub is often the result.

Most of Florida is affected by periodic, natural fire. The high productivity during the rainy season, the pronounced dry season, frequent lightning, and flammable vegetation all contribute to fires that can be frequent or intense. Florida scrub, occurring on less productive sites than more mesic vegetation, burns relatively infrequently (say every 10-30 years) and intensely. On the other hand, grass-dominated habitats like wiregrass-dominated flatwoods (see Chapter VI) burn more frequently and less intensely. The interplay between fire and vegetation is complex and occurs at multiple spatial scales (see Chapters XI and XII).

Most plants in Florida scrub survive fires, resprouting from below-ground storage organs such as rhizomes. The resprouting is rapid but varies with the species, prior plant size, and season. Types of Florida scrub (e.g., scrubby flatwoods) dominated by resprouting oaks and palmettos rapidly regain their height and density in a few years or decades. A few plants take advantage of the short-term openings (gaps) in the resprouting vegetation to complete their life cycles.

Other plants use dormant seeds to deal with periodic fire. Post-fire germination and growth can be rapid and allow species of lesser competitive abilities to survive in the landscape. The seeds that provide the new start are dormant in the soil and cued to germinate by increased light or fire cues such as smoke or heat.

One of these obligately seeding plants is Highlands scrub hypericum, *Hypericum cumulicola*, the source of much of the data in this book (see Chapters IV-V and VII-XIII). This small herbaceous plant can produce seeds as early as its first year of life (see Chapter VIII). It populates the soil with dormant seeds that can germinate in high numbers after fire, even after many years of quiescence below ground. Plants do particularly well in the years after fire, and populations grow rapidly in size during the first decade after burns. However, plants and populations begin an inexorable decline after this first decade, often disappearing above ground until the next fire (see Chapters XI and XII). Increasing competition with shrubs partially explains this decline.

Hypericum cumulicola occurs mainly in a type of scrub termed Rosemary scrub, named after Florida rosemary (*Ceratiola ericoides).* This shrub is different from oaks and palmettos in that it is killed by fire. Florida rosemary accumulates dormant seeds in the soil, much like *H. cumulicola.* After a fire, its seeds also germinate, but on a delayed schedule relative to herbaceous plants like *H. cumulicola.* This delay gives the smaller plants a head start. But these diminutive individuals are likely to do well only where there are large gaps between the growing Florida rosemary plants. As time marches on, these gaps shrink as Florida rosemary and other shrubs expand their canopies and root systems. With enough time between fires, the gaps become too small to support the less competitive species.

Florida scrub is full of endemic plants and animals, species that are found only in a small area. Many of the rare plants are specialists for rosemary scrub and for gaps (*H. cumulicola* is an example). The biodiversity of this landscape is dependent upon fire and the creation and maintenance of gaps among the dominant shrubs. This relationship is also true for some iconic animals of Florida scrub and other upland ecosystems. Florida scrub jays need bare sand to cache acorns and few pines that harbor predators. Gopher tortoises need open areas to forage on herbaceous plants that are most abundant following fires.

Florida Rosemary scrub occurs as relatively well-defined patches of different sizes within a matrix of wetlands, scrubby flatwoods and flatwoods which have higher moisture availability and are more heavily covered by vegetation (Chapter XI). To colonize other Florida scrub suitable patches, endemic plants and animals in Florida rosemary scrub, need to disperse through large tracts of unsuitable habitat. The spatial arrangement of the scrub patches determines the probability of occurrence of many of these endemics, with large and aggregated patches having the highest probabilities of occupancy (Chapter XII). These endemic species rely on varying compromises between dormancy and dispersal to take advantage of the temporal occurrence of optimal habitats a few years after fire, while also minimizing the consequences of suboptimal habitat development associated with post-fire recovery. Their chances of persistence in the landscape are influenced by spatial heterogeneity and frequency of disturbance.

Conservationists and land managers are well aware of the key role of fire in promoting biodiversity in Florida scrub (and other Florida ecosystems) and apply prescribed fire routinely. Robust debate ensues among land managers about the frequency, intensity, seasonality, and patchiness of burns. For example, patchy burns are thought to be important in allowing the co-existence of species with different optimal requirements for different aspects of the fire regime. Unburned patches are key refugia for species that are tolerant of dense vegetation but lacking the ability to resprout and do not form seed banks to recover from fire.

Unfortunately for the biota, the expanding human population in Florida has eliminated a great deal of natural habitat, fragmented the remaining pieces, and successfully reduced the occurrence of natural fire across the landscape. Remaining habitat fragments, if unmanaged, become overgrown with few suitable areas for many plant and animal species. Diversity decreases and the narrowly specialized organisms are particularly likely to disappear. Some species, like *H. cumulicola*, can persist along edges such as sand roads and fire lanes. However, these anthropogenic habitats are not the same as less disturbed scrub, with effects on the demography of many species (Chapter XIII).

Although only about 15% of Florida scrub on the Lake Wales Ridge has been spared from development, the remaining landscapes, when managed with fire, are supporting a vast array of bacteria, fungi, algae, plants and animals, and a fascinating mixture of adaptations and interactions that has received a great deal of study and will receive continued attention from scientists, conservationists, and nature-lovers.

CHAPTER III

CONNECTING MODELS WITH DATA: A QUICK GLANCE AT PROBABILITY

"Extraordinary claims require extraordinary evidence."
– Carl Sagan

Probability: A framework for model arbitration

If models are simplified descriptions of reality, multiple models might be proposed to describe the same phenomena. So, a valid and pertinent question arises: how do we decide which models to favor? In the book "The Ecological Detective", the authors Ray Hilborn and Marc Mangel (1997) make a strong case for letting data arbitrate between competing models. Fitting models to data allows us to judge their relative merit. This seemingly straightforward task can be more challenging than anticipated, as we will see throughout this book. It requires us to be mindful – ask pertinent questions, perform diagnostic tests, keep track of our assumptions, and, most difficult, acknowledge our ignorance.

The idea of allowing data to judge our models is a powerful and appealing one, but every arbitrator needs some guideline or framework to settle disputes. In this book, the underlying framework used to connect data to models is provided by probabilities.

Although there is still disagreement regarding the most useful definition of probability, it is widely accepted that it relates to how likely an event is to occur. It is defined as the expectation of a particular outcome considering all possible outcomes. A major disagreement among competing frameworks is the basis of this expectation: whether it is obtained from a sample representing outcomes of a long-term (essentially infinite) frequency distribution or if it constitutes a degree of belief that must be updated after new data. The probability of an event is represented by a number between 0 and 1 – where 0 signifies the event is impossible and 1 that it always takes place. When taken as a whole, probabilities provide a way to describe uncertainty. A very instructive account on probabilities can be found in Hilborn and Mangel (1997). Below we discuss some relevant concepts.

Probability of finite events

In order to reach a definition of probability that is applicable to the purposes of this book, let's start by looking at a series of simple Venn logical diagrams representing the probability of finite events.

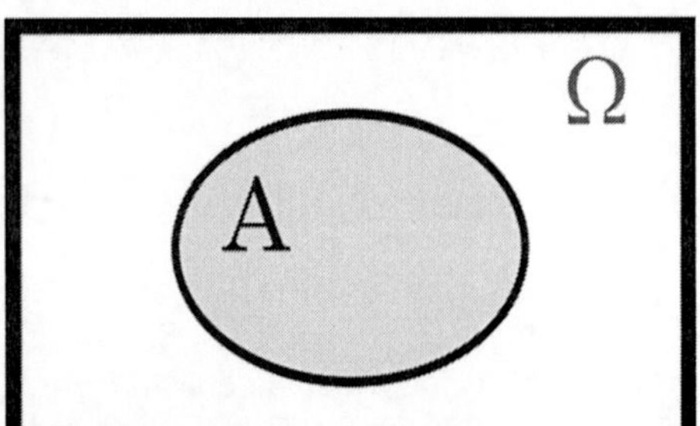

Figure 3.1: Space of event A

If the area of the oval *A* in Figure 3.1 is thought of as representing all instances of event *A*, and the area of the rectangle Ω as representing all instances of any possible event (including event *A*), the probability of *A* can be described as:

$$P(A) = \frac{area\ of\ A}{area\ of\ \Omega} \qquad \text{(Equation 3.1)}$$

We could think, for example, of Ω as all instances of *Hypericum cumulicola* plants and *A* as all instances of adult *H. cumulicola* defined as a plant older than a year. The remaining blank space in rectangle Ω represents all instances of finding recently recruited plants (< one year old). These two events are mutually exclusive. Now that we have a simple definition of probability, we need to turn our attention to another important concept: conditional probability. To do so, we need to look at two sets of events that might occur simultaneously (not mutually exclusive).

In Figure 3.2, there are two overlapping ovals (*A* and *B*), which represent different events that can potentially occur at the same time. At Archbold Biological Station chances of damage by mammals (mostly rabbits and deer) in *H. cumulicola* are very variable (10-90%), decreasing with time-since-fire and increasing with the number of co-occurring conspecifics in the immediate area (Brudvig and Quintana-Ascencio 2003). Imagine that Ω represents all *H. cumulicola* plants in a Florida Rosemary scrub patch. Oval *A* represents all instances of plants with evidence of consumption by mammals in this patch, and oval B all instances of dead plants. In this case, there are 4 possible outcomes: instances of damaged *H. cumulicola* live plants (the non-overlapping section of *A*); instances of intact *H. cumulicola* plants that died (the non-overlapping section of *B*); instances of damaged dead plants (the overlap between both ovals); and instances of intact live plants (the remaining blank space).

Before turning our attention to conditional probabilities, we need to define joint probabilities, which pertain to the interaction between variables. This probability can be visualized as the area of the overlapping section divided by the total area of the rectangle Ω (Figure 3.2).

$P(A \cap B)$ = the probability of *A* and *B* happening simultaneously

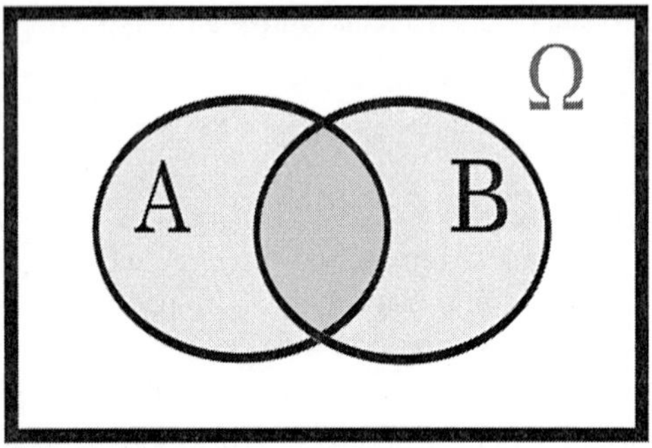

Figure 3.2: Overlapping events

If the probability of a plant dying was independent of damage (i.e., consumed plants have the same mortality as intact plants) we would call these two sets independent events, and $P(A \cap B)$ would simply be the multiplication of *P(A)* and *P(B)*.

You might suspect, however, that the probability of a plant dying is higher for damaged plants. If that were the case, we would not be able to simply multiply *P(A)* and *P(B),* because *P(B)* – the probability of any plant dying – combines information from both damaged and intact plants. So, how do we know if two events are independent? For this we need to look at conditional probabilities.

Let's say event *A* has already taken place (we found a damaged *H. cumulicola*), we want to know the probability that this plant is also dead. The notation for this is *P(B|A),* or the probability of *B* given *A*. The first thing we need to do is recognize that our event space has changed. Any area representing intact plants (both dead and alive) is no longer considered. In other words, we would crop Figure 3.2 and consider only oval *A* and the overlapping section between ovals *B* and *A* (Figure 3.3).

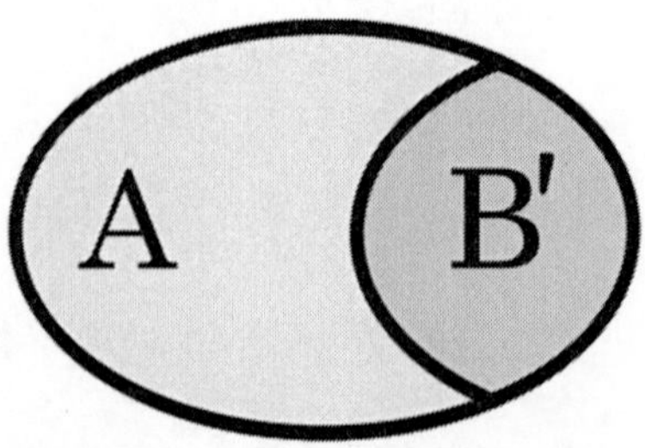

Figure 3.3: Space of *B* given *A*

The probability of *B* given *A*, then, can be visualized as the area of the segment of *B* called *B'* divided by the area of *A*.

$$P(B|A) = \frac{area\ of\ B'}{area\ of\ \mathrm{A}} \qquad \text{(Equation 3.2)}$$

From looking at the sequence of logic diagrams, we can see that, the area of *A* is equivalent to *P(A)* in the original space Ω, and the area of *B'* is equivalent to *P(A ∩ B)*. Therefore, we can rewrite the probability of *B* given *A* as:

$$P(B|A) = \frac{P(A \cap B)}{P(A)} \qquad \text{(Equation 3.3)}$$

Following the same logic, we can define the probability of *A* given *B* (having found it dead, what is the probability that the plant was damaged?). The equation is simply:

$$P(A|B) = \frac{P(B \cap A)}{P(B)} \qquad \text{(Equation 3.4)}$$

Bayes' theorem

As we saw earlier, calculating joint probabilities for non-independent variables is not trivial. So, calculating a conditional probability would still be difficult. We could instead define one conditional probability in terms of the other one. From equation 3.4, then, we can solve the joint probability $P(B \cap A)$

$$P(B \cap A) = P(A|B) \times P(B) \qquad \text{(Equation 3.5)}$$

Since the overlap between *A* and *B* is the same as that between *B* and *A*, *P(B ∩ A)* is the same as *P(A ∩ B)*. So, replacing *P(A ∩ B)* in equation 3.3, we obtain:

$$P(B|A) = \frac{P(A|B) \times P(B)}{P(A)} \qquad \text{(Equation 3.6)}$$

This last equation is called Bayes' theorem. As we will see in a subsequent section, this is a powerful tool that provides a way to revise our beliefs (usually called updating probabilities) given new evidence. Returning to our previous example, calculating the mortality of damaged plants might be difficult because it is hard to estimate how many damaged plants beat the consequences of the herbivory and survived. We could readily get the overall mortality – *P(B)* – from the literature, and conduct a survey to estimate the probability of a plant being damaged – *P(A)* – as well as the probability of a dead plant being damaged *P(A|B)*. Using Bayes' theorem, we could now calculate *P(B|A)*, or the mortality of damaged plants.

Expanding the scope

So far, we have dealt with binary events as realizations of variables that could take one of two values (present or absent, infected or not infected). Random variables, however, may also take more complex forms in which events can have multiple realizations, or even infinite, each with its own probability. In these cases, it may not be practical or even possible to specify each probability individually. To describe these probabilities, then, we can make use of probability distribution functions. The last section of this chapter will present several useful probability distributions and examples of how they can be used to model real data.

Probability distributions are often divided as either discrete or continuous depending on how the variable we want to model is measured. Anything that is counted as integers (such as the number of plants in a population) is considered as discrete, while anything that is measured using units that can be divided into infinitely smaller parts (such as weight) is considered as continuous. It is important to remark that whether a variable is discrete or continuous is contingent on how we perform the measurement rather than the nature of the trait. For instance, we could represent the foliage of a tree as the number of leaves (discrete) or as the total area of the leaves (continuous). The key point to bear in mind is that probability distribution functions are themselves models used to describe a more complex reality and, therefore, are not set in stone.

In describing discrete random variables, we can use *probability mass functions* (PMF) to compute the probability that the variable takes a particular value (x_i), so that:

$$f(x_i) = P(X = x_i) \qquad \text{(Equation 3.7)}$$

The actual form of the function $f(x_i)$ will change depending on the discrete distribution being described, but all PMF share a few characteristics. First, the values of the PMF for each potential x_i must be between 0 and 1, as they represent probabilities. Second, the sum of the probabilities for all realizations of x_i must be equal to 1.

Another way to describe the distribution of a discrete variable is through a *cumulative distribution function* (CDiF) which is the summation of the PMF and computes the probability of a variable being equal to or less than a particular value (x_i), so that:

$$f(x_i) = P(x \leq x_i), \qquad \text{for any } x_i \in R \qquad \text{(Equation 3.8)}$$

To provide a simple example of how the PMF and CDiF relate to each other, let's consider the typical case of a fair six-sided die. The assumed probability of any number in a single roll is given by the PMF:

$$f(x_i) = 1/6 \qquad \text{(Equation 3.9)}$$

which essentially means that *P(1)* = *P(2)* = *P(3)* = *P(4)* = *P(5)* = *P(6)* = 1/6. The corresponding CDiF for this example would be:

$$f(x_i) = \sum_{i=1}^{6} \left(i * (1/6)\right) \qquad \text{(Equation 3.10)}$$

which means that the function will result in the sequence {1/6, 2/6, 3/6, 4/6, 5/6, 6/6}. Individual values of the CDiF cannot be less than 0 or more than 1, but they always remain the same or increase.

When referring to continuous random variables, the probability for a single value is not defined since its mass is infinitely small, but the probability of an interval can be obtained with the integration of a *probability density function* (PDeF).

$$f(x_{interval}) = \int \frac{dF}{dx} \qquad \text{(Equation 3.11)}$$

A *cumulative density function* (CDeF) can also be used to describe continuous distributions of probabilities. They are defined in similar ways as for discrete variables and retain the same properties. The main characteristic of a PDeF is that the total integral must be 1. Just like the CDiF for a discrete variable is obtained through the summation of a PMF, in the case of continuous variables, the CDeF can be obtained through the summation of a PDeF.

$$f(x_i) = P(x \leq x_i), \qquad \text{for any } x_i \in R \qquad \text{(Equation 3.12)}$$

Bayesian inference

In this book we approach statistical inference through a Bayesian framework. In order to describe what this entails, and provide reasons for our choice, we will briefly discuss the approach used in classical hypothesis testing: the frequentist framework.

In frequentist inference, the parameters of a model are treated as fixed, even if unknown, and no probability can be associated with them. Any hypothesized value of a parameter cannot be assigned a probability, so statistical significance is determined by looking at how likely the data are under a null hypothesis (usually stated as the probability of the parameter taking the value of zero).

This notion of likelihood of the data given a parameter value is central to an approach called *maximum likelihood estimation* (MLE), in which we estimate the parameters of a model by finding the values that maximize the likelihood of the data. In other words, if we consider a range of potential values for the parameters of a model, which ones are more likely to produce the data in our sample?

While it is true that MLE is a powerful approach rooted in probability theory, it still fails to assign probability distributions to parameters. In other words, the likelihood function of a parameter is obtained by conditioning the data to different potential values of the parameter and, therefore, it is not a probability distribution function (i.e., it does not integrate to 1). In practical terms, this means that we cannot describe the probability of the hypothesized model parameter value given the data. Yet, this is our stated goal in ecological modeling.

So, how can we go from calculating the probability of the data given a parameter – $P(x|\delta)$ – to calculating the probability of a parameter given the data – $P(\delta|x)$? Luckily for us, Bayes' theorem provides an elegant way to connect these two conditional probabilities, so that:

$$P(\delta|x) = \frac{P(x|\delta) \times P(\delta)}{P(x)} \qquad \text{(Equation 3.13)}$$

This application of Bayes' theorem is known as Bayesian inference, and it presents many advantages. We can start identifying some of those advantages by looking at the different parts of this equation. The term $P(\delta|x)$ is commonly known as the *posterior* and, as stated, represents the probability distribution of the parameter given the data. The term $P(x|\delta)$ is simply known as the *likelihood.* The new term that appears, P(σ), is known as the *prior*, and it represents any knowledge that we possess *a priori* regarding the probability distribution of the parameter. The use of prior information represents one of the main advantages of Bayesian inference and will be treated more carefully in Chapter V. The last term – $P(x)$ – is a normalizing factor that warrants that the posterior is in the range between 0 and 1.

Finding the posterior analytically can only be done in simple cases, for instance by calculating the mean when the shape of the probability distribution of variance can be clearly identified. For most parameters which cannot be solved analytically, there are several ways to estimate their posterior probability distributions. Some rely on assumptions regarding the shape of the posteriors. More flexible procedures that do not require any assumption of distribution usually take much longer and require computation. The estimation of the posterior distribution with these approaches is accomplished using a stochastic process known as Markov Chain Monte Carlo (MCMC) estimation (Hobbs and Hooten 2015). The development of tools to obtain reliable and efficient MCMC estimates is progressing at a fast rate. In the analyses we develop in this book, we depend on the available software STAN (mc-stan.org) to estimate the posterior distributions of our parameters. Discussion on how it works is beyond the scope of this book and we encourage the readers to better inform themselves. There are several excellent sources that explain how these procedures work (for example, Hobbs and Hooten 2015, McElreath 2016).

Modeling randomness: probability distributions

There are many probability distributions that can be used to describe data. Selecting which one to use in our models is a process that requires careful consideration of many questions regarding the nature of the phenomenon and how it is measured. For instance, we should consider what range and types of values the data can take. As mentioned earlier, one of the main ways to classify distributions is by looking at whether they are discrete or continuous. Hilborn and Mangel (1997), Matthiopoulos (2011) and Dietze (2017) provide excellent introductions to these distributions. Here, we briefly present a non-exhaustive list of distributions that we have found to be useful and then show an example of how we might select among them.

Discrete distributions

▶ Binomial

The binomial distribution is used to represent the incidence of an outcome out of a total number of independent trials (called Bernoulli trials), where only two outcomes are possible. We might use it, for instance, to describe the proportion of reproductive plants or the survival probability from one season to the next. The PMF for a binomial distribution is:

$$P(X = k) = \binom{n}{k} p^k (1-p)^{n-k} \quad \text{(Equation 3.14)}$$

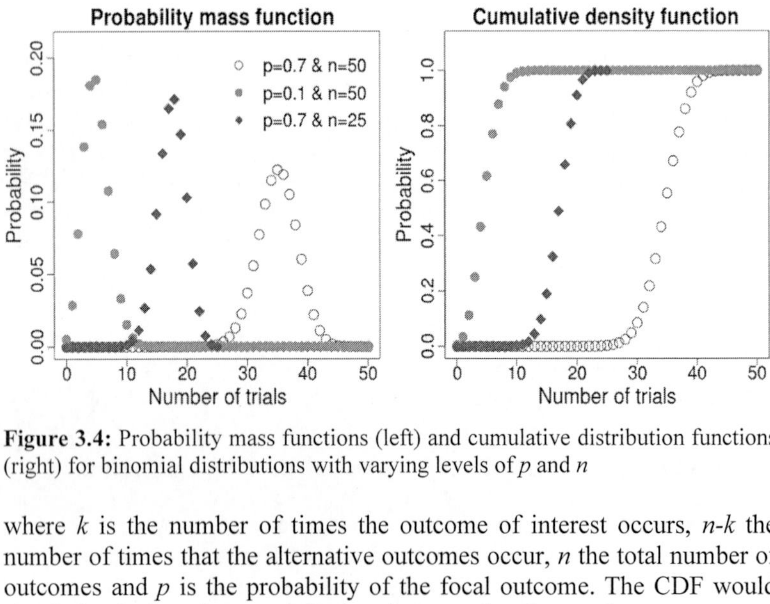

Figure 3.4: Probability mass functions (left) and cumulative distribution functions (right) for binomial distributions with varying levels of p and n

where k is the number of times the outcome of interest occurs, n-k the number of times that the alternative outcomes occur, n the total number of outcomes and p is the probability of the focal outcome. The CDF would simply be the iterative sum of values between 0 and n, so that:

$$P(X \leq k) = \sum_{i=0}^{n} \binom{n}{i} p^i (1-p)^{n-i} \quad \text{(Equation 3.15)}$$

The mean of the binomial distribution is np while its variance is $np(1-p)$. The shape and the location of the binomial functions on the abscissa (x axis), then, will depend on the values of p and n (Figure 3.4).

▶ Poisson

The Poisson distribution is used to represent the probability of a given number of occurrences over a fixed interval (usually of time or space). The Poisson distribution is conceptually related to the binomial, as both deal with the number of occurrences, but is defined by the rate (λ) at which events take place (assumed to be constant). In this way, it is useful to deal with count data. For example, we could use the Poisson distribution to describe the number of plants in a transect or the number of pollinator visits in a day. The PMF of a Poisson distribution is:

$$P(X = x) = \frac{\lambda^x e^{-\lambda}}{x!} \qquad \text{(Equation 3.16)}$$

The *!* symbol refers to the factorial and *e* is the *neperian* or *eulers' number* which is approximately 3.71828. Maor (2011) presents interesting information on *e*. As can be seen in the left plot of Figure 3.5, the mean of the Poisson distribution is the average rate λ. Conveniently, the variance is also λ. This means that, as long as the variance of the count sample is roughly close to its mean, we can describe their probability distribution with a single parameter.

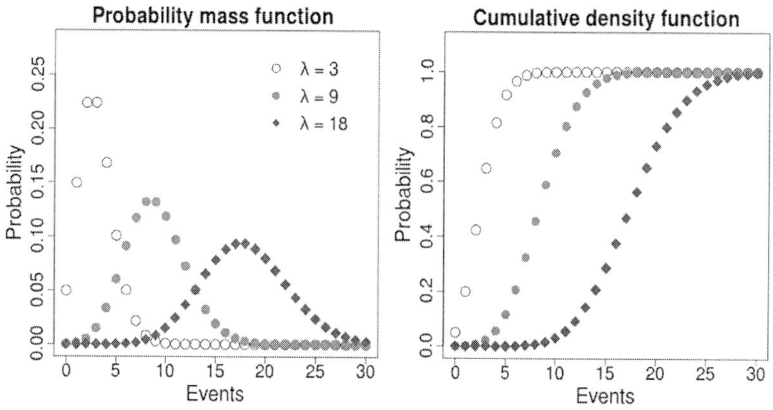

Figure 3.5: Probability mass functions (left) and cumulative distribution functions (right) for Poisson distributions with different λ

► Negative binomial

As the name suggests, the negative binomial distribution is related to the binomial distribution but with the focus flipped on its head. While the binomial distribution looked at the number of occurrences of a particular outcome out of a fixed number of trials, the negative binomial asks how many trials (which can be re-interpreted as how many spatial units, etc.) are necessary to observe a particular outcome a certain number of times. In practice, therefore, this distribution resembles the Poisson, being also useful to deal with count data. However, the mean $\frac{pr}{(1-p)}$ of the negative binomial is different from the variance $\frac{pr}{(1-p)^2}$ making it more appropriate when the mean and the variance of the count data are not similar.

$$P(X = x) = \binom{x+r-1}{x} p^x (1-p)^r \qquad \text{(Equation 3.17)}$$

x is the number of successes, r the number of failures and p is the probability of the successes. As we can see in Figure 3.6 (left), the mean and variance can take different values. Figure 3.6 shows two distributions defined with $r = 3$, the one with $p = 0.7$ has a mean of 2 and a variance of 3.33 while the one with $p = 0.4$ has a mean of 7 and a variance of 23.33.

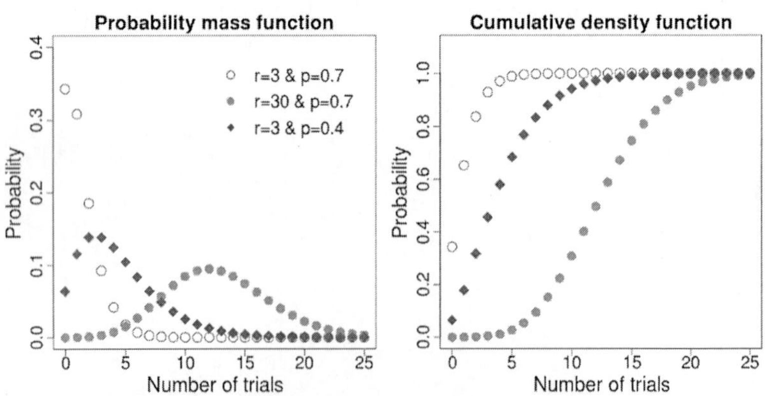

Figure 3.6: Probability mass functions (left) and cumulative distribution functions (right) for negative binomial distributions with different levels of r and p

Continuous distributions

► Uniform

The uniform distribution is sometimes called the rectangular distribution, alluding to its shape (see the left plot in Figure 3.7). Within a given range from "a" to "b", the curve forms a horizontal line representing the equal probability of any value (defined as the inverse of the difference between b and a). The probability of values outside of the range is zero. The PDF of this distribution is, therefore, the piecewise function:

$$P(X=x)=\begin{cases} 0 & for\, x<a \\ \frac{1}{b-a} & for\, a\leqslant x\leqslant b \\ 0 & for\, x>b \end{cases} \qquad \text{(Equation 3.18)}$$

As you might expect, the CDF is also a piecewise function, but defined so that the probability for any value lower than "a" is zero and for any value higher than "b" is one. Within the range, the probability increases linearly.

The mean of the uniform distribution is equal to $\frac{1}{2}(a+b)$, while the variance is $\frac{1}{12}(b-a)^2$. Biological phenomena are almost always, if not always, heterogeneous. It may be hard, then, to imagine how the uniform distribution might be useful in describing patterns of nature. In fact, we do not envision the use of this distribution to attempt to model the posterior probability for any biological variable. Instead, we use it as a statement of ignorance within our models. It is a way of acknowledging that we know very little beyond the potential range of values for our observations. This is useful in some circumstances, such as giving all hypothesized parameter values equal prior probability in Bayesian inference or selecting randomly from multiple options.

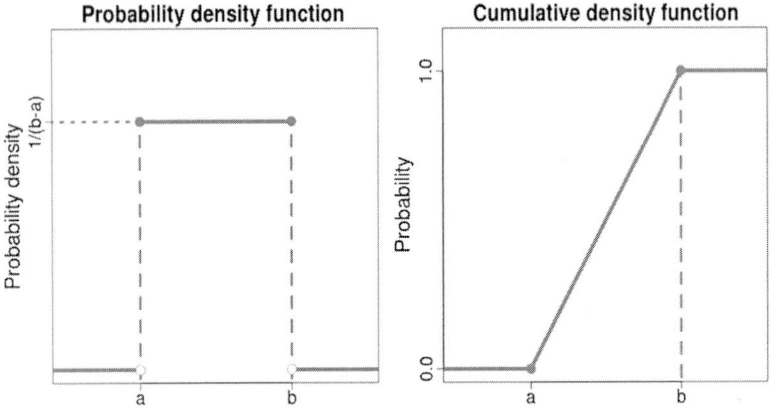

Figure 3.7: Probability mass functions (left) and cumulative distribution functions (right) for a uniform distribution defined by a range [*a*, *b*]

▶ Gaussian

The Gaussian distribution is also known as the normal distribution. The accumulation of random deviations around a central value generates a Gaussian distribution. This distribution is ideal to summarize the uncertainty of phenomena integrating a collection of independent random values from sets of multiple distributions including some other distributions such as binomial, Poisson and negative binomial.

The Gaussian distribution is well suited to representing samples of length and weight and other continuous variables that result from the influence of multiple relatively independent drivers on each individual expression, such as genetic, environmental, and ontogenetic factors. Its PDF is defined as:

$$P(X = x) = \frac{1}{\sigma\sqrt{2\pi}} e^{-\frac{1}{2}\left(\frac{(x-\mu)}{\sigma}\right)^2} \quad \text{(Equation 3.19)}$$

The normal distribution is defined by two parameters: the mean (μ) and the variance (σ^2) and includes two constants, *e*, the *neperian* number, and π, (*Pi*) that is approximately equal to 3.141592. Beckmann (1971) conveys relevant and interesting information on π.

The variance does not share the same units as the mean but the standard deviation $\left(\sigma = \sqrt{\sigma^2}\right)$ does. These parameters have clear and useful properties. The mean is at the center of a symmetrical distribution and is equal to the median and the mode. Approximately 95% of the values lie within two standard deviations, encompassing an informative quantile. The term $(x - \mu)$ in the exponent of the function tells us something important about this distribution: it is symmetrical. When x is equal to μ, the term is zero – and therefore the exponent, is zero. As x moves away from μ, on either side, the exponent becomes increasingly more negative. The further x is from μ, the closer the probability is to zero.

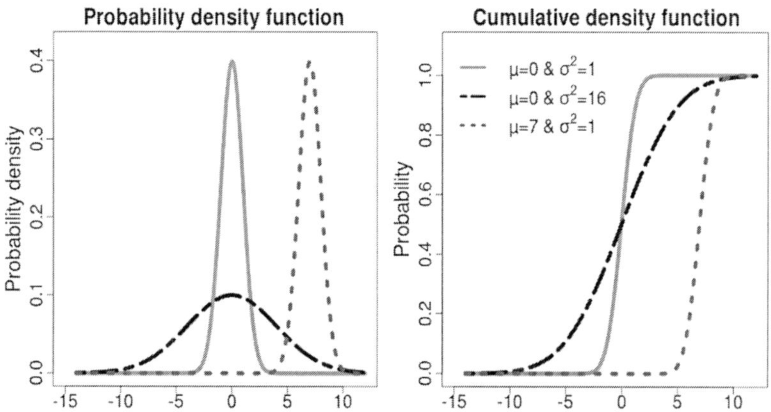

Figure 3.8: Probability density functions (left) and cumulative distribution functions (right) for Gaussian distributions with different means and variances

The Gaussian distribution has been extensively studied, and it is the most used distribution in statistics. This distribution has two useful operations that preserve the form of the distribution. The first one, change of scale, is the operation of multiplying any element X of the distribution by a constant "a" and maintains the shape of the distribution because one unit of X becomes the "a" unit of Y. The second one, shift, is the operation of adding a constant "b" to X which simply moves our random variable X "b" units along the x axis. Thus, using these two operations, any normal random variable can be transformed to a standard normal random variable, having a mean of zero and a variance of one.

▶ Beta

The beta distribution is a flexible option to model continuous variables that may take any value from 0 to 1. It can describe the probability distribution of species turnover (Sorensen index) or percentages when the number of cases is not known. This seemingly confusing statement might be better understood with an example. Let's say we are interested in knowing the proportional area of ground cover occupied by grasses. This can range from 0, no grasses, to 1, all the area is covered by grasses. We might want to know what is the probability that half ($x = 0.5$) of the area is covered by grasses, or that at least 30% of the area is covered with grasses ($x \geq 0.3$). See Warton and Hui (2011) for suggestions on how to analyze this type of data. The PDF of the beta distribution is:

$$P(X = x) = \frac{\Gamma(\alpha+\beta)}{\Gamma(a)\Gamma(\beta)} x^{\alpha-1}(1-x)^{\beta-1}\ , \ for\ 0 < x < 1, \quad \text{(Equation 3.20)}$$

where Γ refers to the Gamma function $\Gamma(\alpha) = (\alpha - 1)!$, and α and β are called shape parameters. The mean of the beta distribution is equal to $\frac{\alpha}{\alpha+\beta}$ and the variance is $\frac{\alpha\beta}{(\alpha+\beta)^2(\alpha+\beta+1)}$.

Given the large diversity of shapes that the beta distribution can achieve (Figure 3.9), we could wonder why we do not always use this instead of others such as those mentioned previously. The main reason is that the interpretation of α and β is not straightforward, making it harder to propose a data-generating process. This distribution represents what we call a conjugate prior to many of the other important distributions – those dealing with successes and failures, such as the binomial and negative binomial. Indeed, if we ignore the first part of the function (which is a normalizing constant), we see terms that resemble those of a binomial distribution.

In simple terms, the beta distribution describes the probability of successes, with α determining rates of success and β rates of failure. You might think of it as a tug of war, with α pulling the distribution to the right and β to the left. In practical terms, using the beta distribution can be convenient to define priors for its conjugate distributions because it helps us avoid more intensive computations. Why? Because we are using a closed-form (having a limited range) prior to calculate a closed-form posterior. Calculating every possible denominator in Bayes' formula takes time, so limiting the prior to the range of interest is helpful.

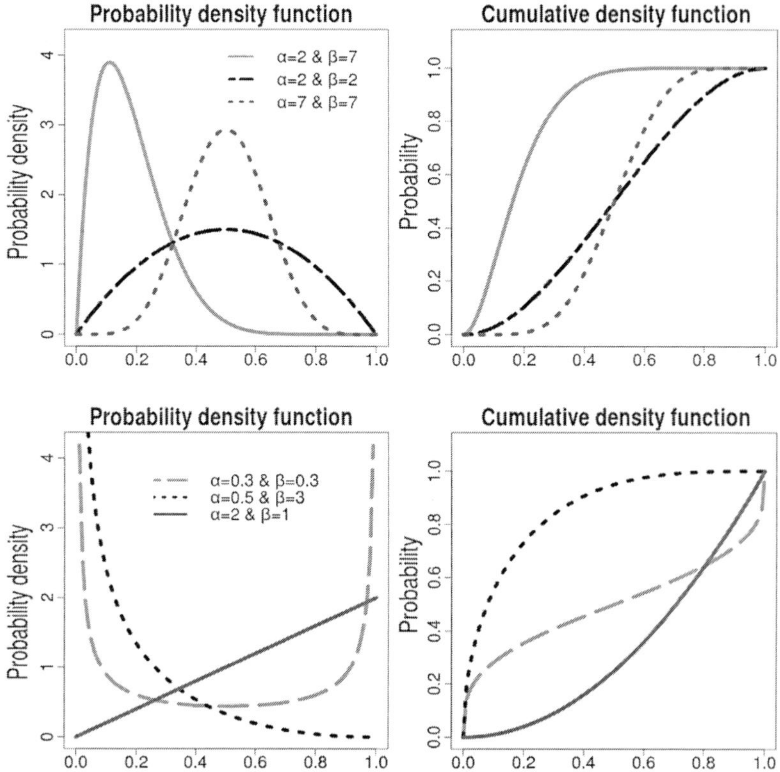

Figure 3.9: Probability density functions (left) and cumulative distribution functions (right) for beta distributions with different α and β

▶ Gamma

The gamma distribution is also a flexible continuous probability distribution. In Bayesian statistics, it is used as a conjugate prior of the precision (inverse of the variance) of the normal variance and of various types of inverse rate parameters such as λ of a Poisson distribution. The shape of this distribution can be characterized with two shape parameters α and β.

$$f(x) = \frac{\beta^{\alpha} x^{\alpha-1} e^{-\beta x}}{\Gamma(\alpha)}, \quad \text{for x, } \alpha \text{ and } \beta > 0, \qquad \text{(Equation 3.21)}$$

where Γ refers to the Gamma function $\Gamma(\alpha) = (\alpha - 1)!$ and e is the neperian number. The mean of this distribution is calculated as $\frac{\alpha}{\beta}$, and the variance as $\frac{\alpha}{\beta^2}$. With these two parameters, many potential distributions can be obtained. In fact, both the exponential and chi-squared distributions are special cases of the gamma distribution. Examples of this distribution are in Figure 3.10.

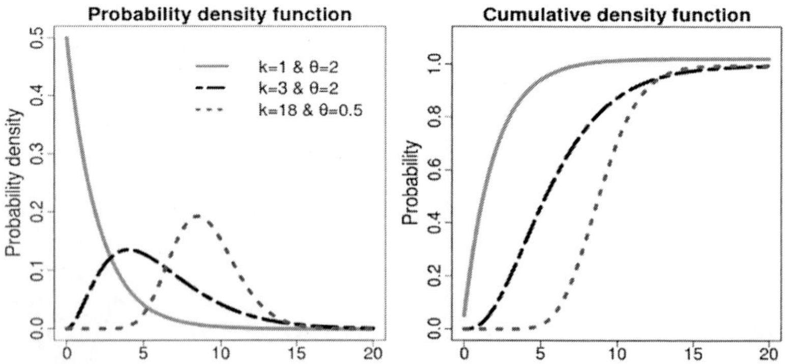

Figure 3.10: Probability density (left) and cumulative distribution functions (right) for gamma distributions with different αs and βs

▶ Log-normal

Defined simply, a log-normal distribution is one in which the logarithm of the variable of interest follows a Gaussian (or normal) distribution. Just like with Gaussian distributions, the log-normal distribution is well suited to represent variability that arises from multiple independent drivers deviating observations from a central tendency. An important difference between the two is that log-normal variability represents a multiplicative effect of such drivers, while Gaussian variability is the result of additive effects.

Due to the properties of logarithms, variables that are log-normally distributed must possess a couple of characteristics. First, the variable must take values greater than zero. Remember, logarithms are defined only for $0 < x \leq \infty$. Second, a log-normally distributed function will be skewed to the right (positive skew), since there needs to be the same number of values between 0 and 1 as between 1 and $+\infty$. So, log-normal distributions are very useful to describe phenomena with low mean values but large variances that cannot be negative, such as fecundity, species abundance or the distribution of nutrients in the environment. The mean of the log-normal distribution is

$exp\left(\mu + \frac{\sigma^2}{2}\right)$, while the variance is $(exp(\sigma^2) - 1)exp(2\mu + \sigma^2)$. The PDF of a log-normal distribution is:

$$P(X = x) = \frac{1}{x\sigma\sqrt{2\pi}} e^{-\frac{1}{2}\left(\frac{(ln(x)-\mu)}{\sigma}\right)^2}, \; for\, x > 0, \qquad \text{(Equation 3.22)}$$

where μ is the mean of the log-normal distribution in the natural scale and the shape of the distribution is determined by σ. The results of varying the shape parameter can be seen in Figure 3.11.

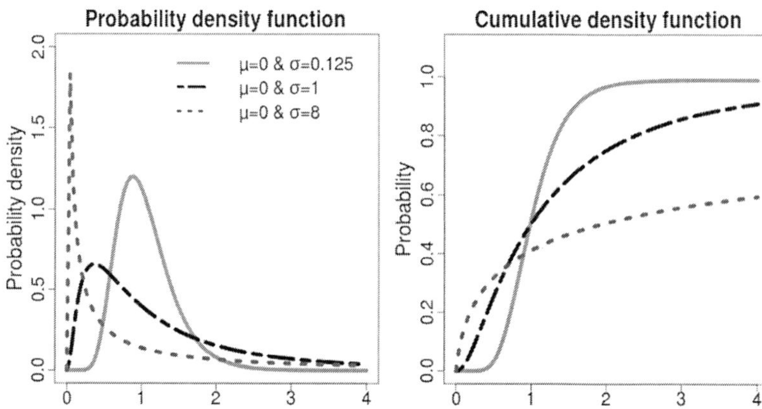

Figure 3.11: Probability density functions (left) and cumulative distribution functions (right) for log normal distributions with means of zero and varying σ

Further considerations

When selecting a distribution to represent the probability of some data, we must make some considerations regarding the assumptions we are willing to make. We must carefully consider how the data are generated and how they are measured. The first question we should ask, and maybe the most obvious, is whether the data are better considered as discrete or continuous. Second, it would be helpful to consider if our measurements might be restricted to certain ranges. A third important question we might want to ponder over regards how much we know about the potential values the data could take, what are the sources of variability and how they might make certain values more common than others.

Let's conclude this chapter by looking at an example of how these considerations can help to select a probability distribution to describe some real-world data. For this purpose, we use height (in cm) data of the rare

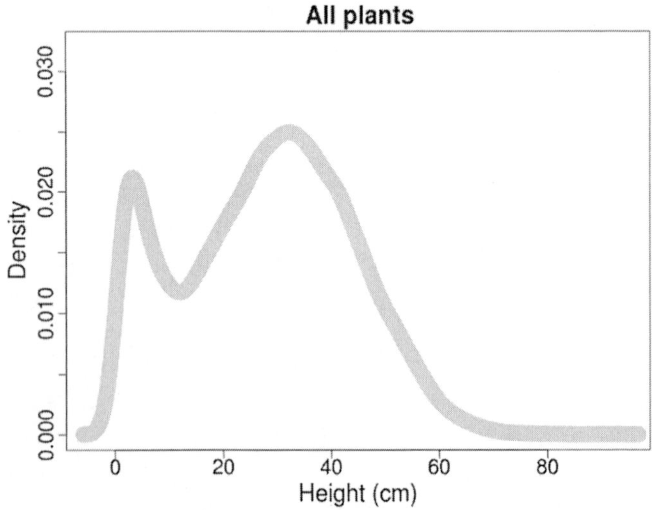

Figure 3.12: Observed distribution of heights for all *Hypericum cumulicola* plants

Florida scrub plant *Hypericum cumulicola* collected by Quintana-Ascencio and colleagues (2003, 2018, 2019) across 14 populations at Archbold Biological Station (an ongoing study with ~40000 cases of ~16000 plants, collected annually since 1994). Here, we explore height data collected between 1994 and 2014. We model these and other related data more deeply in the following chapters. For now, we focus on representing the overall shapes of their distribution.

The first thing that becomes evident is that the overall height data seem to have two separate modes: one around 5 cm and one around 33 cm (Figure 3.12). Our knowledge of the focal species lets us suspect that the source for this pattern is the different behavior of yearlings (new recruits) when compared to adult plants.

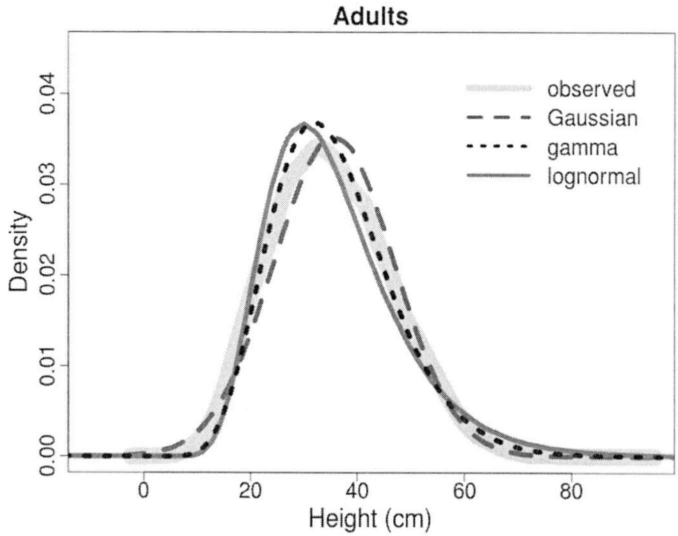

Figure 3.13: *Distribution of Hypericum cumulicola* adult height

Even though we might explore ways to represent multi-modal distributions, we feel justified in exploring each group separately. If we subset adults, then, we see a different pattern (Figure 3.13).

The Gaussian distribution seems to do a fairly good job at representing the overall shape of the data regarding adult height. The mode might be slightly off (as the observed data have a small positive skewness), but the results might be acceptable. However, we should also consider that the Gaussian distribution has no range restriction and predicted heights could take negative values (however unlikely it may seem from the graph). To solve this inconsistency with the allowable range for heights, we might look at the log-normal distribution which can only take positive values. This distribution also does an acceptable job at representing the data, but it exaggerates the skewness. The gamma distribution does a slightly better job at representing both the shape of the data and the range of values that the variable can take. However, its interpretation might be more difficult. Now, considering again our original observation regarding yearlings vs. adults, we should explore if these same observations hold for yearlings.

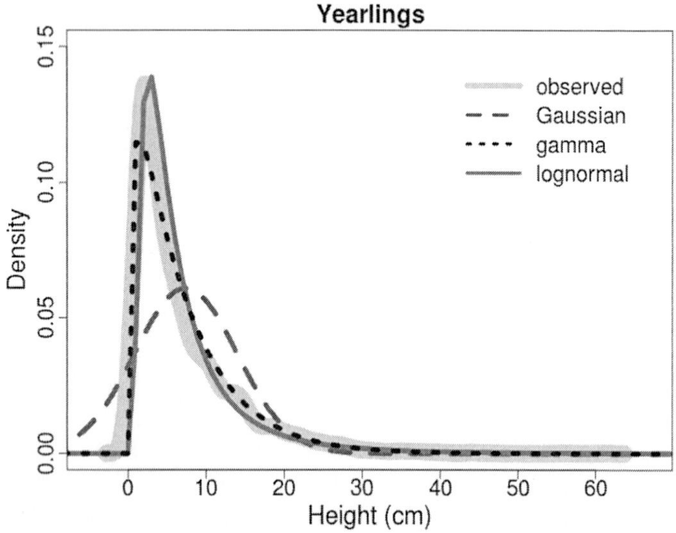

Figure 3.14: *Hypericum cumulicola* yearling height

The distribution of yearling heights is positively skewed, with most plants being between 1 and 10 cm (Figure 3.14). In this case, the Gaussian distribution fails both in recovering the overall shape of the data and in restricting predictions to a reasonable range (now probable values below zero are clearer). Both the log-normal and gamma distributions do a much better job, with the log-normal distribution seemingly doing better.

The main takeaway from this example is that it is always convenient to analyze the data weighting different options. A visual assessment of the data is always recommended.

CHAPTER IV

BAYESIAN APPROACHES: THE AVERAGE HEIGHT OF A RARE PLANT

"Science progresses one funeral at a time."
– Max Planck

Relevant notions

In his Critique of Judgment, Immanuel Kant famously claimed that there could never be "a Newton of a blade of grass." This statement referred specifically to the genesis of life from inanimate matter, and, in this sense, his prediction still holds. However, when we talk about the origin of life forms, Charles Darwin has come incredibly close to deserving the moniker. His ideas provided the framework for the modern study of life – a framework rooted in variation and change.

We collect samples because we recognize variation as being inherent to nature and wish to describe it. In simple terms, a sample is a collection of things or some information about those things. To summarize these collections, we can derive useful properties of the said sample and, after some assumptions, of other similarly collected samples from the same population and even properties of the whole population (here we refer to the set of all the elements constituting a "statistical population"). These properties are called **descriptive statistics** and include, among others: the central value, the spread of the sample, and the shape of the distributions of the sample and the population. In ecological studies, we often want to generalize the properties of the collected sample in a way that can also be expected in other similarly collected samples. This can be achieved through **inferential statistics**. As discussed in the previous chapter, there are two widely used approaches in inferential statistics: **frequentist statistics** and **Bayesian statistics**. In general, inferential statistics can be used to (1) predict properties of other samples, and (2) evaluate relative information from different models that predict these properties.

The two statistical frameworks differ in their conception of probability. The frequentist approach defines probability as fixed, long-term frequency. This means that, if we could sample to infinity, we would be able to find the exact value of a parameter. The Bayesian framework, on the other hand, defines probability as a degree of belief regarding potential values of a parameter. Such a belief is updated with new information. We suggest that the Bayesian view is, therefore, better suited to deal with a field characterized by variation and change – especially when considering our sampling limitations.

Frequentist approaches do not allow for hypotheses or unknown parameter values to be represented as probabilities. Bayesian approaches characterize the uncertainty of competing hypotheses as probabilities. So, for instance, since the frequentist framework considers the central value of a population (the population mean) and its spread (the population variance) to be fixed, it does not estimate probabilities for different hypothesized values of these parameters. It can only estimate the population mean and variance as single values that may or may not be included in the range of the confidence interval of the model specified given the data observed. The Bayesian framework considers that the population mean (or any parameter) is not a fixed value, thus representing the plausibility of different values probabilistically and allowing subsequent collections or other sources of information to update its distribution.

Practical insights

The estimation of averages and variances is an important step in many ecological analyses, such as the demographic modeling performed throughout this book. In this chapter you will be introduced to *R* and *Stan* functions which will allow you to reach these goals. Simultaneously, you will be confronted with the two alternative analysis frameworks – frequentist and Bayesian – and their application in estimating averages. As we have already mentioned, these two approaches represent different ways to evaluate information. We offer you an example that will help you decide their relative advantages to make inferences from data. We will use the data introduced in the previous chapter – heights of *Hypericum cumulicola* collected by Quintana-Ascencio and colleagues (2003, 2018, 2019).

Imagine it's 2014 and you have been tasked with measuring the height of *H. cumulicola* at a site close to Archbold Biological Station. You might decide to take a sample of 10 plants, or you might take a sample of 100 plants. We will create these fictional samples by randomly selecting 10 individuals and 100 individuals from the 2014 field season. We will use each sample set to construct a frequentist model and two Bayesian models – one with what is known as *diffuse priors* and the other with *informed priors*. We are convinced that there are no diffuse priors that are completely non-informative. However, they are used as a statement of lack of prior information, and often they are built as a quasi-flat distribution assuming equal or quasi-equal probabilities for any possible value of the parameters. The objective is to characterize the mean and variation of the *H. cumulicola* height and obtain an estimate of their uncertainty, and to explore how each model deals with different sample sizes.

This example dataset and all data used in the subsequent chapters can be found on the website of the book. The example file for this chapter contains data on the standing height in centimeters for each sampled plant in Archbold Biological Station measured annually between 1994 and 2014. It also has information allowing us to calculate the mean and variance in height for each of the fourteen sample populations by year or in total. The variables and settings of the model are specified in an *R* (R Core Team 2019) script that is sent to *Stan* (Stan Development Team 2019) for processing and results are retrieved by *R* for visualization – in other words, for the analysis the user needs only to interface with *R* but code needs to be written in both programs.

To see how the *R* script works, let's take a look at the code. For this part we will concentrate on the upper portion of the script. The first line in the program *rm(list=ls())* allows you to clear the memory. This is good practice when you start a new script, as it helps to avoid confusion with other currently used scripts. The second and third lines are used to obtain and set the directory that will be used in the program (Box 4.1). You will need to modify these lines to adapt it to your directory pathway or preference for data access. The function *read.csv("file name.csv", header=T)* makes the data available for computation. The next line filters the plant height data to only include plants older than their first census *(Hc_data$stage !="sg")* and allocates them to a vector *"Height_data"*. Symbols *"!="* in *R* mean "not equal to". As mentioned in the previous chapter, plants younger than their first annual census have different demographic properties than older plants. We will explore some of these differences in greater depth later in the book.

Box 4.1: Code for data retrieval and visualization

```
# Function to read the data as a csv file

Hc_data <-read.csv("Hypericum_cumulicola.csv", header=T)
Height_data <- Hc_data$ht_init[Hc_data$stage !="sg"]

# Function to build a histogram

hist(Height_data,100,main = "Histogram of Hypericum
cumulicola height (cm)")
```

Calculating the average and variance of a very large sample

Before we continue with our example, let's briefly discuss the process of characterizing the average and variance of a large sample. To this end, we will use the 17,552 observations recorded between 1994 and 2014. It is always a good idea to start with a visual assessment of the data, checking its structure and distribution. We use the function *hist()* to create a histogram of the observed adult plant heights, and the function *abline()* to add a bold black line for the location of the overall mean of these data. We also add distributions of the expected normal, beta and gamma distributions generated using the observed mean and variances. Your plot should look like the one in Figure 4.1 where the data seem to follow a left-truncated, quasi-normal distribution with more than the expected frequencies for small heights. This deviation between data and modeled distributions may have resulted from demographic variation or unintentional biased sampling.

You might be surprised to find that the assumed normal distribution (a continuous black line; obtained with the function *rnorm()*) predicts some probability for non-positive values. Regardless of the metaphysical debate that it might elicit, negative height is not something that we observe in the field, so you might be doubtful of this distribution. Alternative specifications assuming a beta distribution (a dashed gray line; obtained with the function *rbeta()*) and a gamma distribution (a dotted gray line; obtained with the function *rgamma()*) are limited to positive height values, but the location of their modes (the peak of the distribution) does not coincide with the mean of these distributions nor do the beta or gamma expected distributions overlap with the observed data as well as the normal distribution. We should remark that all three distributions have the same mean and variance as the observed data – the difference in their modes is a result of the skewedness of each distribution. As it was more consistent globally, we are inclined to continue assuming a normal distribution for the analysis of these data.

There are alternative ways to obtain an average and a variance in *R* (Box 4.2). Here we first use the function *mean()* to get an overall plant height mean of 31.82 cm and the function *var()* to get an overall variance of 179.98 (standard deviation = 13.42). We also obtain the value of the standard error of the plant height mean (SE = 0.10 cm) by calculating the square root of the population variance and dividing it by the square root of the degrees of freedom (*n*-1) and rounding the result to four digits. The function *lm()* can accomplish these calculations in fewer steps but the terminology can be confusing (output in Box 4.3).

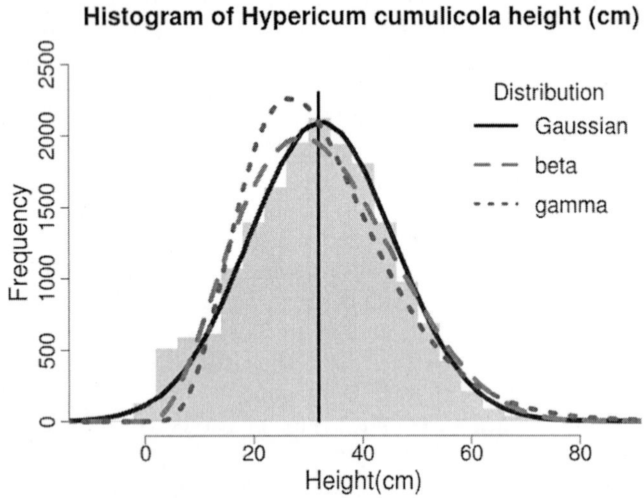

Figure 4.1: Histogram of the individual height of *Hypericum cumulicola* (1994-2014, n=17552), and densities of the different distributions used to describe the pattern, all of them with the same mean (vertical line) and variance

Box 4.2: Code to calculate the mean and variance of the overall data

```
mean(Height_data)
var(Height_data)
round(sqrt(var(Height_data))/sqrt(length(Height_data)-
1),4)

summary(lm(Height_data ~1))
```

Box 4.3: Example of the *R* output of the function *lm*

```
lm(formula = Height_data ~ 1)

Residuals:
  Min      1Q   Median    3Q     Max
-31.32  -8.82   0.18   9.18  59.18

Coefficients:
            Estimate  Std. Error  t value Pr(>|t|)
(Intercept)  31.8196     0.1013   314.2   <2e-16 ***
```

```
Signif. codes:  0 '***' 0.001 '**' 0.01 '*' 0.05 '.'
0.1 ' ' 1

Residual standard error: 13.42 on 17551 degrees of
freedom
```

🕶 Glasses On 🕶

Unbiased estimates of population variance and degrees of freedom

The **population variance**, also known as the second central moment, is defined by:

$$\sigma^2 = \frac{1}{N}\sum_{i=1}^{N}(x_i - m)^2, \qquad \text{(Equation 4.1)}$$

where m is the assumed, known and fixed (in frequentist approaches) population mean and N includes all the members in the statistical population. This formula produces a biased estimate of the population variance when calculated using a random collection of size n, a portion of sampled elements from the N population elements with an *a priori* unknown mean (the mean needs to be estimated). This biased estimate occurs because of the loss of one degree of freedom. When we estimate the **unknown mean** we reduce the amount of independent information, leaving only n - 1 independent elements (degrees of freedom). The unbiased estimator of the population variance (the sample variance) is:

$$s^2 = \frac{1}{n-1}\sum_{i=1}^{n}(x_i - m_s)^2, \qquad \text{(Equation 4.2)}$$

where s^2 is the sample variance, m_s is the estimated sample mean and n is the number of elements in the sample. The amount of bias from the first equation rapidly decreases with the size of n.

Preparing the data for analysis and obtaining prior information

To help assess how well each model performs in estimating the mean height of *H. cumulicola* for different sample sizes, we will first peek into the long-term data to provide a frame of reference. While observations made in a shorter period of time might be biased by a common factor (e.g., a drought), we believe data collected in 14 populations for over two decades are a reasonable reference for comparison. We will, therefore, calculate the mean for all 17,552 observations collected between 1994 and 2014. We will use this information to evaluate the relative efficiency of our models to estimate the means and variances from a sample of 10 plants and from a sample of 100 plants. As already mentioned, we will compare the estimates obtained from the collected samples using frequentist and Bayesian approaches with diffuse and informed priors.

There are a few variables common to all the models that we need to create. Looking at the code, you can see that we define the sample size (10 or 100 plants) as *n*. We also define a vector *x* of length *n* to contain the randomly selected observations. For the Bayesian model with informed **priors**, however, we need to define a few more variables. In Chapter V, we discuss the use of prior information in greater depth. For this example, we simply use observations collected between 1994 and 1997 to construct the prior distribution. These data were published in 2003 by Quintana-Ascencio and colleagues, so they would be available to you as a researcher – in general, previous publications make for great sources of priors.

The variable *pop.mean* is the mean of the means of the heights of plants sorted by population, and the variable *pop.var* is the mean of the variances per population. Figure 4.2 shows the distributions of these two variables. We assume a normal distribution for the informative prior of population means but notice that the distribution of the population variances can never include values smaller than zero (by definition) and is skewed to the right.

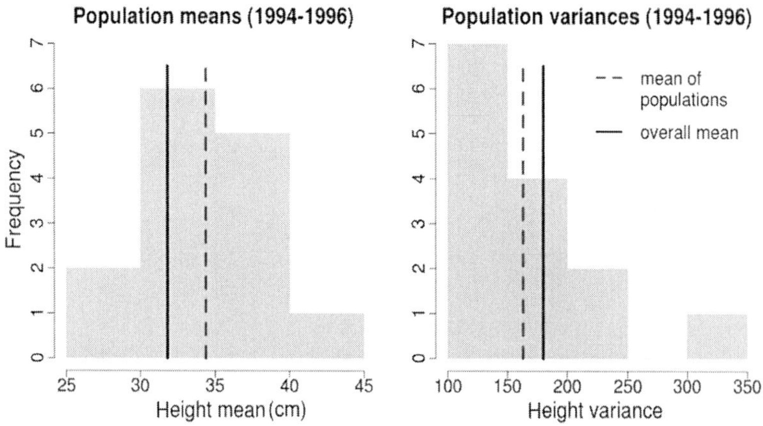

Figure 4.2: Histograms of population level mean and variance of height among 14 populations of *Hypericum cumulicola* (1994-1997). The dashed line represents the mean of population means and population variances. The solid line represents the long-term mean and variance

Log-normal distributions do not include values smaller than zero and are skewed to the right representing a reasonable distribution for the prior of variances. We want to emphasize that this is just a hypothetical scenario. We are not advocating for splitting data to create priors.

Calculation of averages from a small sample (n=10)

For the **frequentist model**, we calculate the mean and variance of the samples employing the formulas described above. For the Bayesian approach, we use functions in the *R* library (rstan) to connect with *STAN*, and in *STAN* we specify the model. The written code is as described below. For the **first Bayesian model**, we define two **diffuse priors**. Notice that the specification of the distributions is not consistent between *R* and *STAN*. The first diffuse prior describes a normal distribution with a mean of zero and a large variance, *normal (0, 100),* and it barely appears at the bottom of its plot (Figure 4.3 left plot). The second one is a diffuse prior for the standard deviation using a Cauchy distribution – *cauchy(0, 1).* This distribution defines an equal likelihood for any value between 0 and 50. Both priors are implying our assumed lack of previous information. In Figures 4.3 and 4.4, the distributions of these diffuse priors are depicted as almost flat horizontal lines. It is difficult imaging, assuming a completely uninformed prior but here, at least, we are giving any reasonable values equal preference.

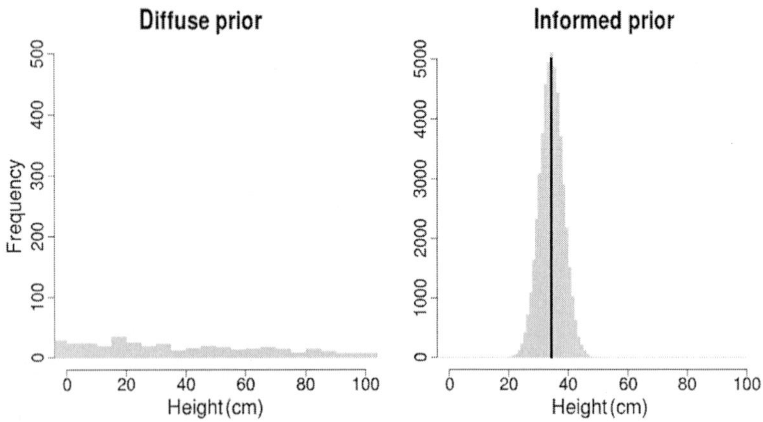

Figure 4.3: Distributions of the suggested priors for the mean. The vertical line is the location of the mean of the means of 14 populations (1994-1997)

For the **second Bayesian model** we use two **informed priors** based on data arranged at the population level as described above. These types of summarized data can be frequently obtained from published sources. In this case we have prior relevant information. We have a mean and a variance from each population collected between 1994 and 1997. The prior that we use for the mean has a normal distribution and is specified as *normal (34.39, 3.93).* The prior for the standard deviation has a log-normal distribution and is specified as *log-normal (2.24, 0.07).* These are our initial beliefs. Compare the distributions of these priors in the right plots in Figure 4.3 and 4.4.

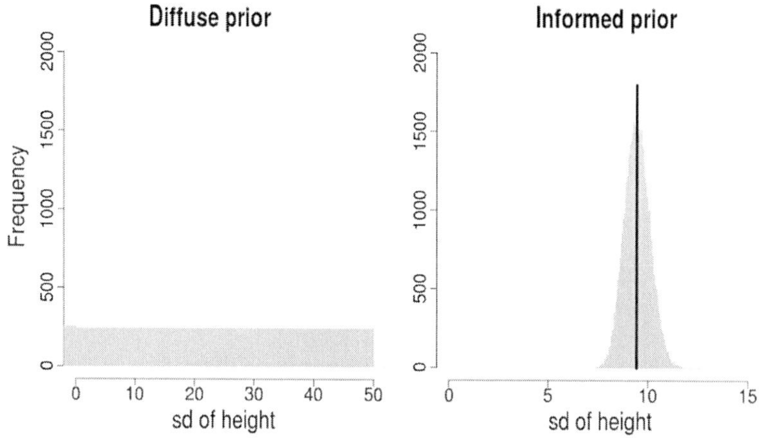

Figure 4.4: Distributions of the suggested priors for the standard deviation. The vertical line is the location of the mean of the standard deviations of 14 populations (1994-1997)

With the written *R* code, we make data available for computation, establish the settings of the run, call the Stan program, check the run record, and retrieve, summarize and plot the posteriors (Box 4.4). In the *Stan* code, we specify the model, declare the variable types, the parameters, the likelihood distribution of the data, the priors for all the parameters and the generated quantities. For information on Bayesian models relevant to ecological studies read McCarthy (2007), Hobbs and Hooten (2015) and Elderd and Miller (2016).

Calling the two programs generates Markov Chain Monte Carlo (MCMC; see Chapter III) chains of values that summarize the distribution of the posteriors – we must also define the number of MCMC chains, the number of iterations per chain, the number of burned-out iterations and the amount of thinning. As the chains start with random and independent values, we must discard the initial portion of the draws to avoid biased information. We also want to minimize correlations between consecutive values and thin the amount of data sampled to build the posterior distribution. Finally, we should visually ensure that there is a good mixing of the chains of each parameter in the model (Figure 4.5) to be confident that the posteriors are robust estimates of the parameters. These trace plots show how the estimation of parameters changes from one iteration to the next – most often the chains are presented in a single plot to help assess whether they overlap well. Another indicator of good mixing among chains is to check if the index $\hat{r}$ (R-hat) is very close to 1.00 (McElreath 2016). Increasing the number of iterations can help the convergence of the model, but it will increase the run time. The *R* function *stan()* determines the procedure for the model and allocates the results to a model object (i.e., Model 4.1) We use the *R* function *print()* to call the results.

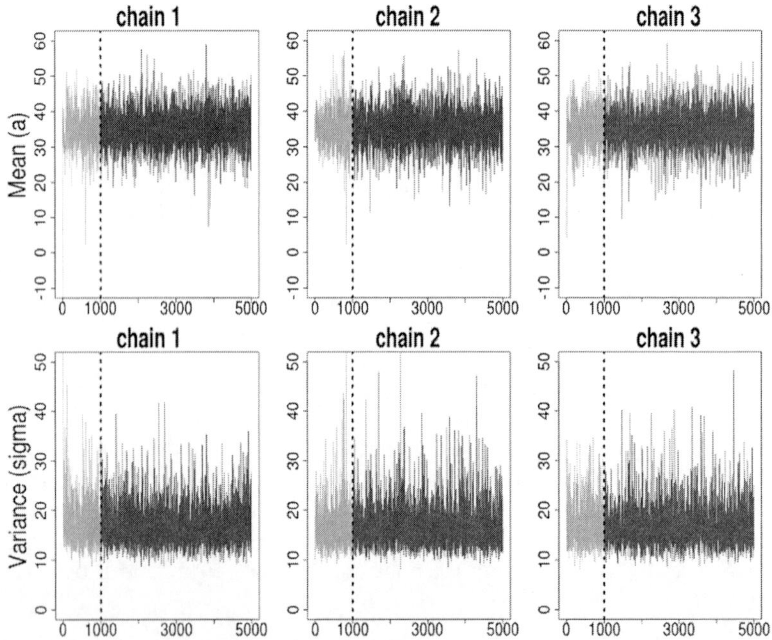

Figure 4.5: Three MCMC chains of the posterior mean and variance of the model with diffuse priors. The area in light gray indicates the burned iterations. Only those in black are used to build the posterior distributions

Box 4.4: Constructing the code to run the analysis. We need to build two sets of code. The first one is an *R* script, including the following:

I. Defining the settings of the run

```
library(rstan) # it calls the library
rstan_options(auto_write = TRUE)
# it saves the compiled instance in our directory
options(mc.cores = parallel::detectCores())
# it runs the chains in parallel
```

II. Gathering the sample

```
n<-10
x <- sample(Hc_data$init_height[Hc_data$time==2014 &
Hc_data$stage != "sg"],n)
stan_data <- list(height=x, N=length(x))
```

III. Setting the parameters monitored

```
params <- c("a","sigma")
```

IV. Defining the MCMC settings

```
ni <- 5000          # number of iterations
nt <- 2             # interval for data collection
nb <- 1000          # number of iterations discarded
nc <- 3             # number of chains
```

V. Defining initial values

```
inits <- lapply(1:nc, function(i) {
  list(a = rnorm(1,0,30), sigma = runif(1,0,15))})
```

VI. Calling for *Stan* from *R* for Model 4.1. It should be replaced with Model 4.2 for the informed model

```
Model_1  <- stan("Chapter_3_difuse.Stan",
                 data = stan_data, init = inits, pars =
params, chains = nc, iter = ni, warmup = nb, thin = nt,
seed = 1, open_progress = FALSE)
```

VII. Plotting and summarizing the posteriors

```
pairs(Model_1,pars=c("a","sigma"))
print(Model_1, digits = 3)
```

The second set of programs consists of two *Stan* scripts. They differ in a couple of lines where the priors are introduced. Notice that the commented lines start with # in *R* and // in *Stan*. Below is the code for Model 4.1.

```
// Data declaration
data{
  int<lower=1> N;
```

```
  real height[N];
}
// Parameter declaration
parameters{
  real a;
  real<lower=0> sigma;
}
// Model declaration including prior specification
model{
  vector[N] mu;
  sigma ~ cauchy( 0 , 1 );
  a ~ normal( 0 , 100 );
  for ( i in 1:N ) {
    mu[i] = a;
  }
  height ~ normal( mu , sigma );
}
// Declaration of generated quantities
generated quantities{
  vector[N] mu;
  real dev;
  dev = 0;
  for ( i in 1:N ) {
    mu[i] = a;
  }
  dev = dev + (-2)*normal_lpdf( height | mu , sigma );
}
```

These are the two lines that change for Model 4.2.

```
  sigma ~ lognormal( 2.245235  , 0.06751135 );
  a ~ normal( 34.39132 , 3.927421 );
```

For an implementation of Model 4.1 for a sample of 10 plants with diffuse priors, we obtained an estimate of the mean (28.83 with SD = 3.9) and sigma (the standard deviation of the variance of the model; mean = 11.51 with SD = 2.9; SD refers to the standard deviation of the parameters estimated from these sampled data). The output also includes the 95% credibility intervals for these parameters. Keep in mind that the results will vary for each random sample. These estimates are slightly different from but commensurate to those obtained with the frequentist approach for the same sample (28.77 and 11.14, respectively for the mean and sigma).

Table 4.1: Estimated mean and sigma and their standard deviations (SD), and their uncertainty obtained from the three models implemented from the observed data

Source	**Mean**	**SD**	**Sigma**	**SD**	**CI 95% ***
Model estimates from samples in 2014					
Frequentist model	28.77	3.5	11.14		20.8 – 36.7
Bayesian					
with diffuse priors	28.83	3.9	11.51	2.9	21.0 – 36.7
with informative priors	30.90	2.4	9.60	0.6	26.2 – 35.7
Estimates from reference data					
Priors (1994-1997)	34.39	3.9	9.44	1.1	
Overall (1994-2014)	31.82	0.1	13.42		

* Credibility interval for Bayesian models, confidence interval for frequentist models

The implementation of Model 4.2 for the same sample of 10 observations uses an informed prior based on the mean and standard deviation of the population height means and sigma (the standard deviation) from 14 populations in the period 1994-1997. The obtained mean and standard deviation (and their SDs) are different (30.9, 2.4 and 9.60, 0.6) from those calculated from the model with diffuse priors. The value of the mean of the model with informed priors was closer to the overall mean (31.82) and its standard deviation was smaller than for those of the frequentist or Bayesian model with diffuse priors. Summaries of the realizations of the two models are in Table 4.1. The precise values of each realization will change with the sample used (see more examples in Figures 4.7 and 4.8).

This realization of the Bayesian model with informed priors had closer values to the mean of the populations and much narrower 95% credible intervals (26.2-35.7) than those estimated with the uninformed Bayesian model (21.0-36.7) or the 95% confidence interval of the frequentist approach (20.8-36.7). The comparison between credibility intervals and confidence intervals is not correct since they represent different notions, and we only use this as a heuristic tool. Figure 4.6 shows the distributions of the informed prior, the posterior of the model with diffuse priors and the posterior of the model with informative priors for the mean height of a sample of ten plants in 2014. In the first Bayesian model, because the diffuse prior has very little information, the distribution of the posterior is predominantly characterized by the information from the data (and is consistent with the frequentist approach). In the second Bayesian model, our updated belief, the posterior of the model, can be interpreted as a compromise between the distribution of the prior and the likelihood of the data. In this Bayesian model with informed priors, the posterior distribution is more consistent with the long-term average (the black line in Figure 4.6). We encourage you to repeat the whole process but with a sample of 100 *(size <-100)* to explore how the sample size *n* affects the outcomes. Below, we provide the results of 20 runs of this example using samples of n=10 and n=100 (Figures 4.7 and 4.8).

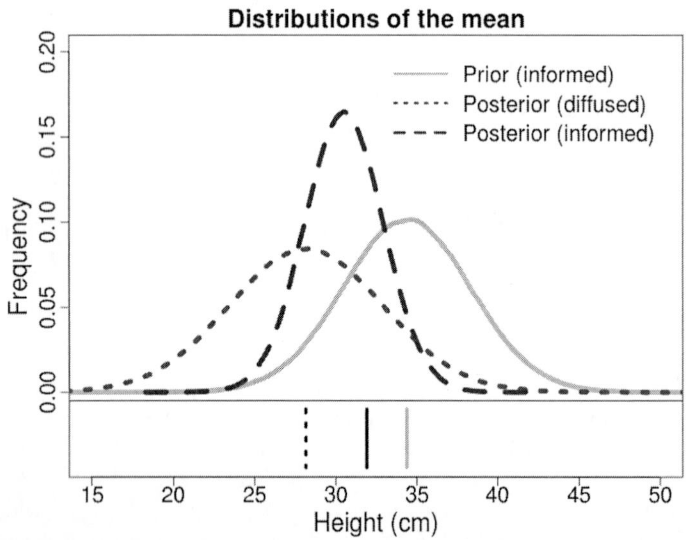

Figure 4.6: Distributions of the informed prior and the posteriors for the diffuse and informed models for a sample of 10. Bottom: mean populations in the period 1994-1997 (solid gray line), the overall mean from 1994 to 2014 (solid black line), and the frequentist mean for the same sample (dashed gray line)

-- --

🕶 Glasses On 🕶

The credibility interval and confidence interval are different. Bayesian posteriors are probability distributions and credibility intervals define the range of values included by different quantiles of these probability distributions. In a different way, the frequentist level represents the long-term frequency of possible confidence intervals containing the assumed fixed "true" value of the statistic. For example, it is expected that 85% of an infinite number of 85% confidence intervals will contain the fixed mean. Crucially, an 85% confidence level does not mean that for a given realized interval there is an 85% probability that the population parameter lies within the interval.

In *R* the range of a confidence interval is calculated as:

*mean(x) - qt(0.85, n-1)*sqrt(var(x)/n)*
*mean(x) + qt(0.85, n-1)*sqrt(var(x)/n)*

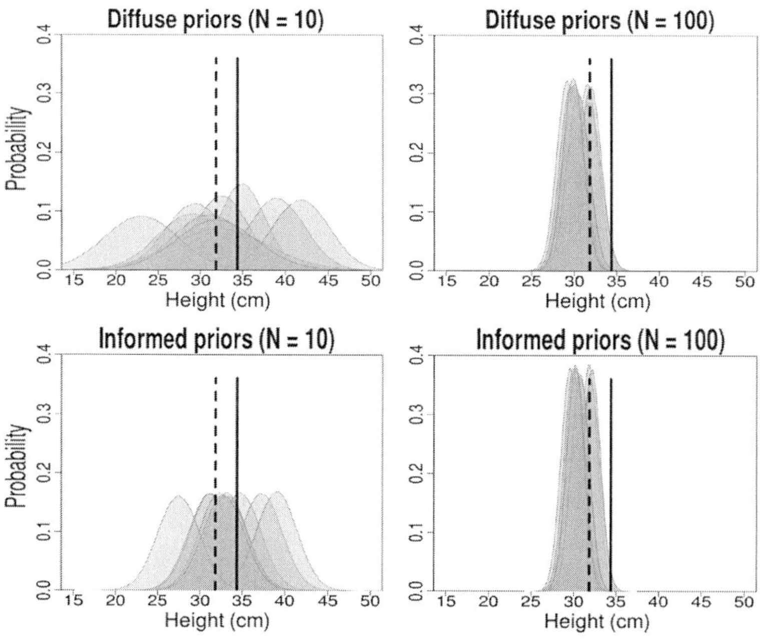

Figure 4.7: Distributions of posteriors of the mean for diffuse and informed models for samples of 10 and 100 plants obtained from plant height data in 2014. The solid black line indicates the position of the average among populations in the period 1994-1996. The dashed black line indicates the position of the overall average (1994-2014)

Final considerations

Figures 4.7 and 4.8 demonstrate that the inclusion of priors allows for more precise and robust estimates. With samples of 10 plants, posterior distributions of models with informed priors are closer to the long-term average and have narrower uncertainty than those from the diffuse models. We also demonstrate that the influence of the prior decreases with sample size. Statistics from models with samples of 100 plants with and without priors are closer to each other. Concerns expressed about the possible misuse of misleading or irrelevant priors have been addressed by Ellison (2004). The potential misuse of information is not exclusive to Bayesian approaches, and frequentist approaches have been the target of similar concerns. We are convinced that naive effects of priors will be reduced and eventually eliminated during normal scientific work as beliefs are updated. Meanwhile, informative priors will contribute to better information and predictions.

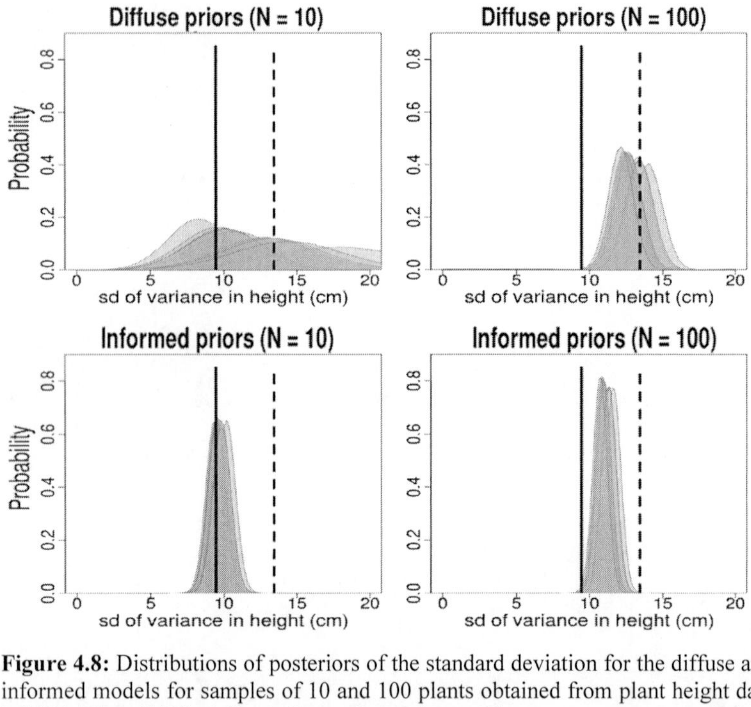

Figure 4.8: Distributions of posteriors of the standard deviation for the diffuse and informed models for samples of 10 and 100 plants obtained from plant height data in 2014. The solid black line indicates the position of the average among populations in the period 1994-1996. The dashed black line indicates the position of the overall average (1994-2014)

Like many other scientists in the 20th and 21st centuries, we were trained and initiated our careers employing frequentist approaches for our statistical analysis; however, during our professional practice we became increasingly disappointed about their inadequate assumptions and, particularly, about their lack of consideration of previous or related information. The lack of training opportunities, scarce source code, and, importantly, limitations on the speed and memory of available computers precluded widespread acceptance and use of Bayesian approaches. All these shortcomings are increasingly reduced, and circumstances are more favorable for taking advantage of Bayesian analysis and avoiding the shortcomings of frequentist approaches. We hope that this book serves as a source to facilitate this transition.

CHAPTER V

THE IMPORTANCE OF ASSUMPTIONS: EVALUATING REPRODUCTIVE OUTPUT

"Your assumptions are your windows on the world. Scrub them off every once in a while, or the light won't come in."
– Isaac Asimov

Relevant notions

In the realm of business administration, most failed plans start with a set of faulty assumptions. This is the premise of an influential *Harvard Business Review* article written by Rita Gunther McGrath and Ian C. MacMillan. There, they warn against the unfounded certainty that arises when assumptions are mistaken for facts. When recognizing uncertainty as unavoidable, the acceptance of assumptions will always be necessary. Good planning maximizes the learning we derive from failures. To achieve this goal, they introduce an interesting and useful concept: the assumption-to-knowledge ratio. The larger the number of assumptions relative to knowledge, the more careful we must be in tracking and testing our assumptions. To this brilliant proposition, we would only add this: facts are never fully understood with certainty. The degree to which we believe we understand a fact, however, should change with new evidence.

Statistical modeling provides one way to formalize this process. When applying statistical models, we consent to a whole set of assumptions. Some people even consider that models are essentially a list of assumptions (Pearl and Mackenzie 2018). While the inference from some models might be robust enough to withstand some deviations from those assumptions, if formulated incorrectly they can lead to inaccurate conclusions.

Assumptions may differ depending on the statistical analysis, and whole books have been written on their importance and how to assess them. To exemplify, we will focus on the assumptions associated with one of the most used statistical tools: regression models. These include:

i. a model specification (i.e., the shape of the expected relationship),
ii. a likelihood distribution of errors (assessed with a probability distribution),
iii. a pattern in the variance of errors (often assumed to be homogeneous),
iv. variation of the explanatory variables (often assumed to be collected without error), and
v. independence of observations.

In simpler terms, fitting a statistical regression model requires us to assume a relationship that is represented by an equation and a likelihood distribution for the uncertainty around the model predictions. There are many ways to specify a relationship between a response variable and one or several predictors.

In the case of linear regression all the terms of the equation are additive, and each term includes a coefficient multiplying a single variable or, if interactions are evaluated, a coefficient multiplying several multiplying variables. Linear regression describes a linear response or one that can be approached as a linear model and whose residuals are assumed as random fluctuations of the observations around that predicted line. Here there is an unfortunate confusion because the word "linear" is used with two different meanings: as a linear "relationship" and as a linear "response." For example, by using logarithmic transformations, exponential and power models can be estimated with linear models. The relationship between the log-transformed response variable and the predictors is linear. Models with polynomial terms of a variable (such as quadratic and cubic models) are also linear models. However, in the natural scale, none of these models have a linear response.

The fluctuations of the data around the predictions of a linear model define the model residuals and are derived from phenomena whose realization is described by a probability normal function and do not necessarily remain constant across the explanatory variable. The emphasis placed on residuals provides an important clue about the best way to evaluate the assumptions of linear models – we look for inconsistent patterns in the variability that remains after fitting the model.

Power and exponential models can also be estimated using non-linear functions which can describe more diverse responses. Statistical non-linear models require the specification of probability distributions to the uncertainty around model predictions. The flexibility of these models to represent more variable functional shapes comes with the cost of increasing complexity and computational time. We concur with other authors in recommending starting with simpler models and then moving toward evaluating more complex models, constantly assessing the benefits and costs of this increasing complexity (Hilborn and Mangel 1997).

For simplicity and mathematical convenience, most models also assume that the data are independent samples from the chosen distributions. Independence among the samples implies that we assume there are no other relevant drivers differentiating the response for certain groups of individuals (sample units) in the sample.

-- --

🕶 Glasses On 🕶

Errors and residuals are two closely related terms. Both refer to the difference between observed values and some other value: the expected or theoretical value in the case of errors and the predicted or estimated value in the case of residuals. More simply, a residual is an observable estimate of the unobservable statistical uncertainty. If the above assumptions on the likelihood of the data of a linear model are met, then the observations will spread around both sides of the predicted values with a pattern across the values of the predictor consistent with the assumed likelihood distribution.

-- --

Practical insights

Demographic studies mostly consist of characterizing the variation of different vital rates – such as growth, survival, maturation, reproduction, dispersal, etc. – based on certain ontogenetic traits, such as size or age, and environmental drivers. In this chapter, we will construct a series of regression models to evaluate an estimate of reproductive output, the change in the number of fruits of *Hypericum cumulicola,* as a function of plant height. These models can be combined with those of other vital rates to build demographic models describing the species' life history and dynamics (see Chapter XIII).

Understanding the patterns of variation in the reproduction of a population is central to predicting its future trends. Reproducing is, after all, a risky affair requiring parents to make some sacrifices. The risk level of each reproductive event is determined by how the individual deals with the challenges imposed by its environment. Thus, reproduction varies among individuals and habitats, and can be altered in response to anthropogenic forces, such as management actions. Aggregating from the individual to the population level can help us to predict changes in abundances and chances of population persistence. In this chapter, we discuss how to verify some model assumptions, how to recognize some of the consequences of their violations, and how to use the insight to improve our decisions.

Estimating reproductive outputs in a way that accounts for environmental, individual, and genetic variability is incredibly challenging. Counting fruits and seeds in plants can be extremely difficult and time-consuming – some plant species, such as velvet grass (*Holcus lanatus*), a European native that is now a pest weed in many regions of the world including the southeastern US, can produce tens of thousands of fruits in one season. The timing of reproduction can also present significant challenges. Some plant species might reproduce during an extended period, requiring recursive sampling, while others might reproduce in a narrow time window or at a time that makes extensive sampling difficult. In the example presented in this chapter, the typical *Hypericum cumulicola* individual produces hundreds of small fruits during the hot and humid summers of Central Florida. No matter how hard we try, counting the minuscule structures in the scorching, bright sand of the scrub will lead to some level of measurement error.

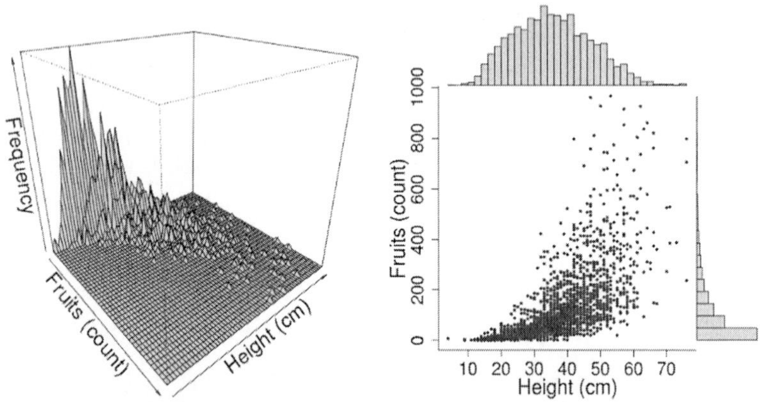

Figure 5.1: The number of fruits as a function of plant height for *H. cumulicola.* In the right plate, fruit data were truncated to less than 1000 fruits. The observed maximum is approximately 6000 fruits

These difficulties often lead to data that are limited, incomplete, associated with many sources of uncertainty and only partially represent the phenomena of interest. During analysis, we are forced to ignore many sources of variation. While statistical models allow us to derive meaning from these data, they do so within a simplified framework that requires the acceptance of a series of assumptions. The assumptions we are willing to accept affect the quality and value of our models and inferences.

We will concentrate on the change in the number of fruits as a function of standing plant height in 14 populations of *Hypericum cumulicola* in Florida rosemary scrub patches at Archbold Biological Station. The data were collected during the Augusts of 1994, 1995 and 1996. Here, we will concentrate only on reproductive individuals. Figure 5.1 depicts the frequency distribution of these data and the relationship between the focal variables.

These plots indicate that intermediate sized reproductive individuals occurred more frequently in the sample than smaller or larger reproductive individuals. Additionally, the graph shows that variation in the numbers of fruits increased with height, potentially violating the frequent assumption of homoscedasticity of linear models. This general pattern is common among many plant species, and it will be important to keep in mind as we explore different formulations for the model.

Model building

With the aim of describing the change in the number of fruits with height, we can start naively by analyzing these data using a linear model of the relationship between these two variables with a normal likelihood distribution for the data. Here, we also assume no previous relevant information and propose diffuse priors for the coefficients and the variance, centered on zero and with broad variances. Model 5.1 summarized below (Box 5.1), indicates that the number of fruits increases with height. If we accept the assumptions as valid, we could state that Model 5.1 explained approximately 24% of the observed variation.

The underlying statistical Model 5.1 is:

$$Number\ of fruits = \mathrm{a} + \mathrm{b} * height \quad Likelihood\ N(\mu, \sigma)$$

Box 5.1: Simplified implementation in *R* and *Stan* for Model 5.1

```
model{
    vector[N] mu;
    sigma ~ uniform( 0 , 400 );
// prior for variance
    b ~ normal( 0 , 100 );
// prior of coefficient for height
    a ~ normal( 0 , 500 );
//prior of intercept
    for ( i in 1:N ) {
        mu[i] = a + b * height[i];
// model specification
    }
    fruits ~ normal( mu , sigma );
//likelihood
}
```

The results obtained from running three MCMC chains with 1000 sampling iterations are in Table 5.1 and Figures 5.2 and 5.3. The 95% credibility intervals include specified quantiles of the posterior estimates for each coefficient. The Rhat is an index of the consistency of the estimates across the chains performed by the MCMC (as seen in the previous chapter, values close to 1 are desired).

Table 5.1: Mean, standard deviation, 95% credibility intervals (CI) and Rhats of the coefficients in Model 5.1 and of its uncertainty as the standard deviation of the residual error (sigma)

Coefficient	Mean	Std Dev	lower CI	upper CI	Rhat
a	-269.4	17.1	-297.2	-243.1	1.00
b	11.2	0.4	10.5	11.9	1.00
sigma	240.8	3.7	234.8	246.5	1.00

If we accept its assumptions, Model 5.1 indicates a change of approximately 10-12 (coefficient b) fruits per unit increase in height – this is the slope of the line.

An inspection of the chains of the MCMC procedure (Figure 5.2) indicates that they were well mixed and consistent, which suggests no problems with the estimation of these parameters. In Chapter IV, we presented each chain for each parameter separately to showcase the results of the MCMC process. Each zig-zagging line represented the path taken by each chain from iteration to iteration – these are the values used to calculate the posterior distribution. In this version, all of the MCMC chains for each parameter are overlapped – at the edges, you might see the darker shades from the other chains. If the parameter estimation is robust, however, the amount of overlap makes it hard to see the other chains. This version is more consistent with how trace plots are usually presented than the one in the previous chapter (in the package *RStan*, the function *traceplots()* produces a similar plot).

A pair plot of the individual coefficient estimates obtained in each iteration (Figure 5.3) can also be a very helpful visual tool. The density plots in the main diagonal show the posterior distribution of each coefficient (a, b, and sigma). The scatterplots show the correlation between any given two coefficients. For instance, the upper-right panel shows the correlation between “sigma” (in the x-axis) and “a” (in the y-axis). Each dot represents one iteration of the MCMC procedure. The panels showing only numbers represent the Pearson’s correlation index between 2 coefficients.

The scatterplot in the upper-center panel demonstrates an interesting and frequently overlooked consequence of the specification of many linear models: the correlation between intercept and slope. In general, highly correlated coefficients can be worrisome. This particular occurrence, however, is the result of a simple modeling decision.

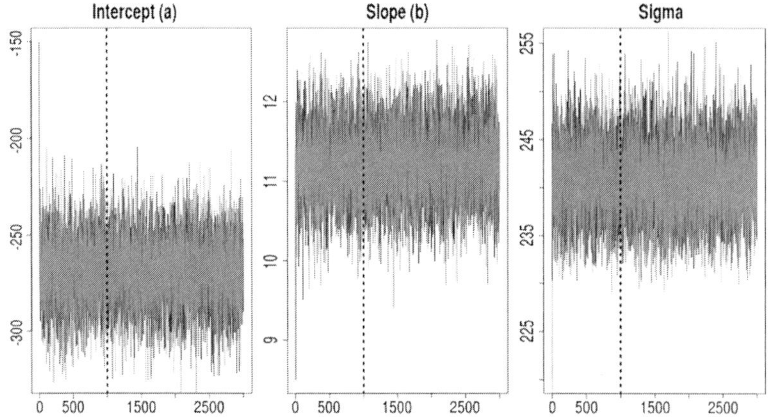

Figure 5.2: MCMC chains of the posterior parameters of Model 5.1. The area to the left of the discontinuous line indicates the burned iterations. Only those on the right of the line are used to build the posterior distributions

-- --

🕶 Glasses On 🕶

The strong correlation between the intercept and the slope is not an inherent characteristic of linear models but a consequence of model formulations that place the intercept away from the average value of the predictor. When models are instead defined with a mean-centered predictor, i.e., when the intercept represents the prediction for the mean value of x, the correlation disappears. If we take the linear formula y = β0 + β1 x + ε, the correlation for the estimates of β0 and β1 would be defined as follows:

$$Corr\left(\hat{\beta}_0, \hat{\beta}_1\right) = \frac{-\sum_{i=1}^{n} x_i}{\sqrt{n}\sqrt{\sum_{i=1}^{n} x_i^2}} \qquad \text{(Equation 5.1)}$$

From this we can see that, when we use a mean-centered predictor, Σx will be 0 and the correlation will disappear. You can test this on your own by running a new model which is formulated similarly to the previous model, but with mean-centered height as the predictor variable. To create this new variable, simply subtract the mean of x from each value (see the Chapter IV script on the website for the solution).

-- --

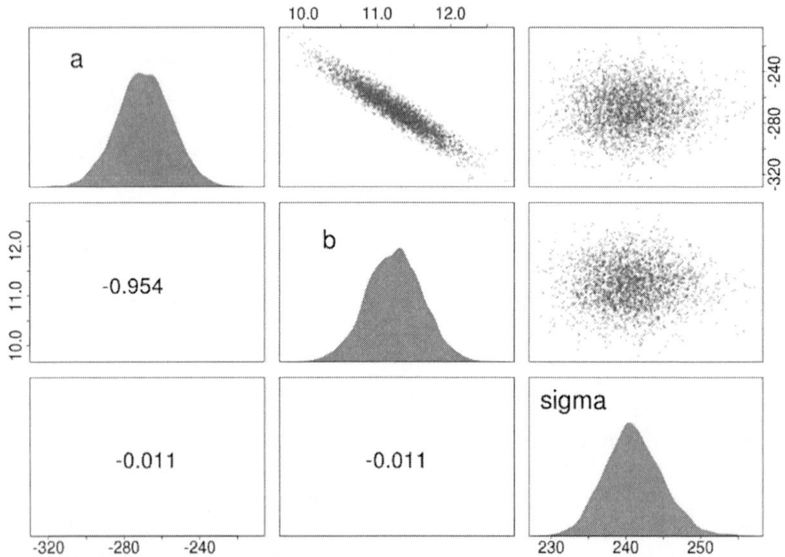

Figure 5.3: Plots of the output of Model 5.1, assessing the variation of the number of fruits as a linear function of plant height for *H. cumulicola* measured at Archbold Biological Station. The plots in the diagonal show the posterior distributions of the parameters and the association among their realizations is indicated numerically (below the diagonal) and as plots (above the diagonal)

A plot of the association between these two variables (Figure 5.4) confirms that the number of fruits consistently increases with height. Simultaneously, we recognize that two of the assumptions made when building Model 5.1 do not seem to have been met. First, the rate of increase in the number of fruits does not seem to be constant, as this linear model implies. Second, the amount of uncertainty in the number of fruits increases with height.

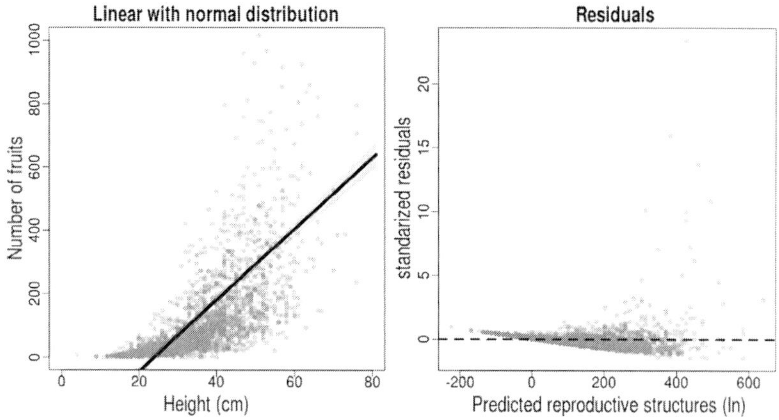

Figure 5.4: Left: Plot of the number of fruits as a function of plant height for *H. cumulicola* measured during 1994, 1995, and 1996 at Archbold Biological Station. The line is the linear model 5.1 (y = -269 + 11x). The polygon is the 95% credibility interval of its mean prediction. The ordinate axis was truncated at 1000. Right: Standardized residuals of Model 5.1

A further inspection of the plot of Model 5.1 (Figure 5.4, left plot) reveals the non-linear shape of the relationship between height and the number of fruits. These observations should make us suspicious of the inferences we can derive from Model 5.1. Particularly worrisome were the predicted negative number of fruits for plants smaller than 24 cm and the pattern of the residuals for Model 5.1 (Figure 5.4, right plot). Further, comparisons of distributions of the observed number of fruits for plants with specified height values against distributions generated using the most likely parameters for the same data for three common distributions with lower limits equal to or higher than zero, indicate a better match with negative binomial distribution and log-normal distribution than Poisson distribution. The frequency distribution of the observed data for plants with heights of 33 and 53 cm is an example in Figure 5.5. For the data with more extreme heights, log-normal distribution seems to better follow the shape of the data.

Residuals – the deviations of the observed data from the predictions of a model – provide important information on the quality of the model. We use standardized residuals to facilitate the comparison across different models since they control for non-constant variance. The formula to calculate them is:

$$\text{Standardized residuals} = \left(\frac{residuals}{sd(residuals)}\right) \qquad \text{(equation 5.2)}$$

Given that we assumed the likelihood of the data having a normal distribution with a constant variance across the values of height, we are also concerned that the residuals of Model 5.1 do not spread evenly around the mean prediction of fruits for each height.

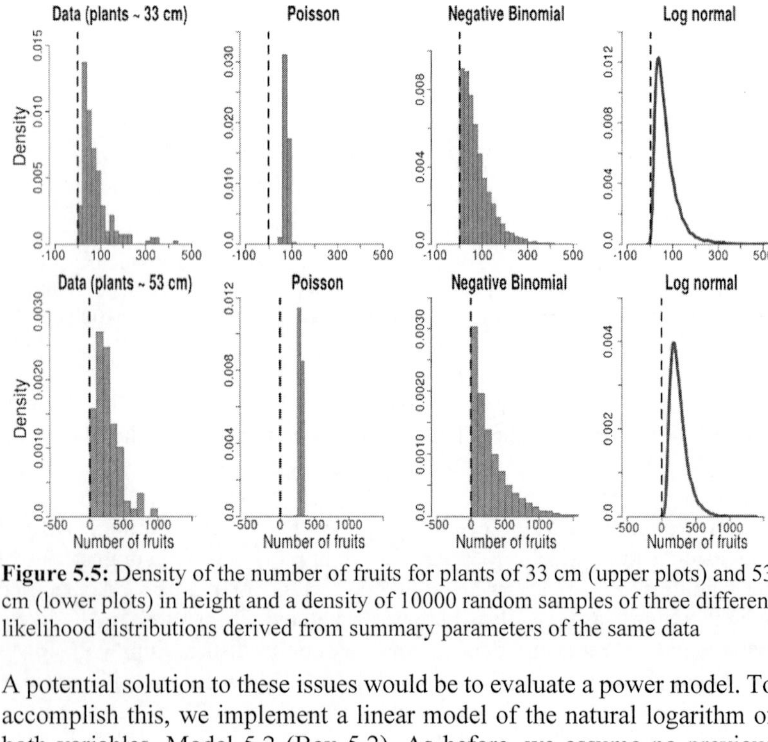

Figure 5.5: Density of the number of fruits for plants of 33 cm (upper plots) and 53 cm (lower plots) in height and a density of 10000 random samples of three different likelihood distributions derived from summary parameters of the same data

A potential solution to these issues would be to evaluate a power model. To accomplish this, we implement a linear model of the natural logarithm of both variables, Model 5.2 (Box 5.2). As before, we assume no previous relevant information and propose diffuse priors for the coefficients and the variance, centered on zero and with broad variances. The underlying statistical Model 5.2 is:

$$\ln(Number\ of\ fruits) = a + \text{b} * \ln(height) \quad Likelihood\ N(\mu, \sigma)$$

Box 5.2: Simplified implementation in *R* and *Stan* for Model 5.2

```
model{
    vector[N] mu;
    sigma ~ uniform( 0 , 10 );
// prior for variance
    b ~ normal( 0 , 10);
// prior of coefficient for height
    a ~ normal( 0 , 100);
// prior of intercept
    for ( i in 1:N ) {
        mu[i] = a + b * ln(height[i]);
// model specification
    }
    ln(fruits) ~ normal( mu , sigma );
// likelihood
}
```

The results obtained from running three MCMC chains with 1000 sampling iterations are in Table 5.2. Model 5.2 characterizes the data much better than Model 5.1 (Figure 5.5), and the more even distribution of the residuals reveals a more reliable model (Figure 5.6). Model 5.2 indicates an approximate 3-unit increase in the logarithmically transformed number of fruits per unit of logarithmically transformed plant height (both using natural-base logarithms). This suggests a change that is faster than linear in the number of fruits as height increases (as an approximately 3 power of height). Model 5.2 has the additional advantage that avoids predicting negative fruits for any size, included the smallest ones; for example, for plants of 1 cm height: exp(-7.1) = 0.0008 fruits.

Table 5.2: Mean, standard deviation and 95% credibility intervals of Model 5.2

Coefficient	Mean	Std Dev	lower CI	upper CI	Rhat
a	-7.1	0.18	-7.4	-6.8	1.00
b	3.2	0.05	3.1	3.2	1.00
sigma	0.8	0.01	0.8	0.8	1.00

When we logarithmically transformed the data, we changed the model's hypothesized shape of the relationship: in this case from a linear to a power model. The estimation of the parameters of a power model varies depending on whether we use logarithms or non-linear models and the different likelihood distributions to approximate the variation on the data. This occurs because it affects our assumptions on the distribution of the variation in fruits for plants of a given height. For the previous power model, we used logarithmic transformed data (natural base) and assumed a normal distribution for the likelihood of the logarithmically transformed reproductive values at each height. An alternative approach is to use a non-linear formulation. Below, we approximate the solution of two non-linear models assuming (i) a normal likelihood, and (ii) a log-normal likelihood for the reproductive values at each height.

The underlying formulations of these two models are:

$$y = \mathrm{a} * height^{\mathrm{b}} \;\; Likelihood\ N(\mu, \sigma_i) \qquad \text{Model 5.3}$$

$$y = \mathrm{a} * height^{\mathrm{b}} \;\; Likelihood\ lnN(\mu, \sigma_i) \qquad \text{Model 5.4}$$

Box 5.3: Simplified implementation in *R* and *Stan* for Model 5.3

```
model{
    vector[N] mu;
    sigma ~ uniform( 0 , 400 );
// prior for variance
    b ~ normal( 0 , 10 );
// prior of coefficient for height
    a ~ normal( 0 , 10 );
// prior of intercept
    for ( i in 1:N ) {
        mu[i] = a * height[i]^b;
// model specification
    }
    fruits ~ normal( mu , sigma );
// likelihood
}
```

Table 5.3: Mean, standard deviation, and 95% credibility intervals for Model 5.3

Coefficient	Mean	Std Dev	lower CI	upper CI	Rhat
a	0.001	0.0004	0.0003	0.0012	1.00
b	3.30	0.11	3.15	3.49	1.00
sigma	232.30	3.56	226.68	237.99	1.00

Box 5.4: Simplified implementation in *R* and *Stan* for Model 5.4

```
model{
    vector[N] mu;
    sigma ~ uniform( 0 , 100 );
// prior for variance
    b ~ normal( 0 , 1 );
// prior of coefficient for height
    a ~ normal( 0 , 1 );
// prior of intercept
    for ( i in 1:N ) {
        mu[i] = a * height[i]^b;
// model specification
    }
    fruits ~ lognormal( mu , sigma );
// likelihood }
```

Table 5.4: Mean, standard deviation and 95% credibility intervals for Model 5.4

Coefficient	Mean	Std Dev	lower CI	upper CI	Rhat
a	0.23	0.012	0.211	0.250	1.00
b	0.80	0.014	0.781	0.826	1.00
sigma	0.84	0.013	0.817	0.817	1.00

The simplified implementation of Model 5.3 in *R* and *Stan* is in Box 5.2, and that of Model 5.4 is in Box 5.3. The results obtained from running three MCMC chains with 1000 sampling iterations are in tables 5.3 and 5.4.

A graphical comparison of the shapes of the responses for the three power models is in Figure 5.6 and that of their residuals is in Figure 5.7. We cannot use Akaike information criteria (see Chapter VII) to compare these models because they do not assume the same likelihood and their dependent variables are in different scales (Burhman and Anderson 2002). Here, we use a visual comparison between the distribution of the data and the predictions of the models to decide among them.

Before we decide which model to favor there is an additional consideration. The number of fruits is usually seen as count data. A potential criticism of the models presented so far is that counts are best modeled using either the Poisson or negative binomial probability distribution. We believe that the selection of probability distributions should be done with careful consideration of what measurements mean. In this case, we consider the number of fruits to be a measure of reproductive effort. When we use centimeters to represent height, we are using a discrete unit to represent continuous phenomena. In doing so we accept there is measurement error due to limitations of the unit. Likewise, here we contend that fruits represent an imperfect unit of reproductive effort, and thus can be represented with continuous distributions.

Nonetheless, we will listen to the criticism and explore a linear model using a negative binomial likelihood, Model 5.5 (Box 5.4). We will compare the predictions to the other models we have developed to decide if such a change was warranted. The underlying formulation of Model 5.5 is:

$$Number\ of fruits = \mathrm{a} + \mathrm{b} * height \quad Likelihood\ NB(\mu, \mathrm{scale})$$

Box 5.5: Simplified implementation in *R* and *Stan* for Model 5.5

```
model{
     vector[N] pbar;
     scale ~ cauchy( 0 , 2 );
// prior for the scale
     b ~ normal( 0 , 10 );
// prior of coefficient for height
     a ~ normal( 0 , 10 );
// prior of intercept
     for ( i in 1:N ) {
          pbar[i] = a + b * height[i];
// model specification
          pbar[i] = exp(pbar[i]);
}
     fruits ~ neg_binomial_2(pbar,scale);
// likelihood
}
```

Instead of AIC (or WAIC; see Chapter VII), we use four other criteria to decide which is the most useful model: (1) a graphic evaluation allowing a comparison of the location of the posterior mean of each model against the distribution of the observed data across the range of the predictor variable (Figure 5.6), (2) the distribution of the residuals around the model prediction (Figure 5.7), (3) the flexibility of the specifications of these models to include other drivers of fruit production, and (4) the location of the mean of the observed data (Figure 5.7). We use the standardized residuals in the logarithm scale to facilitate the comparison among models.

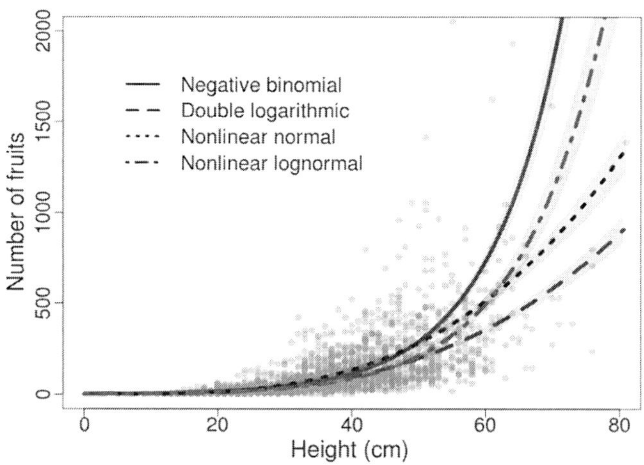

Figure 5.6: Plot of the number of fruits as a function of plant height and the mean model posterior prediction of four models. Model 5.2 uses logarithm transformations in both axes (y = exp(-7.1)*$x^{3.2}$). Model 5.3 is a non-linear model with a normal likelihood (y = exp(-6.0)*$x^{3.0}$). Model 5.4 is a non-linear model with a log-normal likelihood (y = exp(0.23)*$x^{0.8}$) and Model 5.5 is a linear model with a negative binomial likelihood and logarithm link (y = exp(1.9 + 0.09*x). The ordinate axis was truncated in 2000

These four models predict similar increases in the number of fruits with height for most plants, but diverge for plants >40 cm. Models 5.2 (log-log, with a normal likelihood) and 5.4 (non-linear with a log-normal likelihood) predict similar increases in the number of fruits with height for plants smaller than 50 cm, but then diverge. Similarly, Models 5.3 (non-linear with a normal likelihood) and Model 5.5 (the GLM with a negative binomial likelihood and a logarithm link) predict similar increases in the number of fruits with height for plants smaller than 50 cm, but then diverge.

The divergence after 40 cm relates to the skewness of the data better described by the models with a log-normal likelihood. The divergence after 50 cm can be explained by the relative effect of reduced sample sizes, as plants are taller, on the predictors of the different models, that are more sensitive in Models 5.3 and 5.5. The results obtained from running three MCMC chains with 1000 sampling iterations for Model 5.5 are given in Table 5.5.

Table 5.5: Mean, standard deviation and 95% credibility intervals of Model 5.5

Coefficient	Mean	Std Dev	lower CI	upper CI	Rhat
a	1.09	0.059	0.999	1.185	1.00
b	0.09	0.002	0.089	0.094	1.00
scale	1.80	0.055	1.707	1.880	1.00

Notice that the estimated average parameters and their mean predictions were different among the three power models and the generalized linear negative binomial model (Figure 5.6), while the shapes of their responses are similar. They differ as to how fast the number of fruits increases with height, mostly among large plants.

Residuals of Model 5.2 (log-log) deviate uniformly and spread consistently across the entire range of the data (Figure 5.7). Except for the smallest and largest plants, most residuals of power Models 5.3 and 5.4 and the generalized linear model with negative binomial errors, Model 5.5, spread consistently. Residual values of Model 5.3, and especially Model 5.1 were biased (Figures 5.4 and 5.7). Residuals for values of plants with extreme values of height and smaller sample sizes were biased in Model 5.4 and Model 5.5 (they curved at the edges), probably because of the influence of outliers. Figure 5.8 shows the distribution of the number of fruits for plants of different sizes, between ~13 and ~58 cm in height. We used a wider (3-unit) interval for each class to increase the number of observations available for comparison. Overall, Model 5.2 was the most consistent with the distribution of the observed data and their means (logarithmically transformed) by height. Surprisingly, the negative binomial model was not as informative as other models.

Based on this information, we favored Model 5.2. There remains the criticism that the number of fruits is a discontinuous variable. However, conceptually, we consider the number of fruits, a discontinuous variable, to be an imperfect estimate of reproductive effort, a continuous variable, and prefer the log-normal probability distribution as the likelihood for these data. It better represents the long tails of the distribution of the number of fruits when split by plant height and is less sensitive to the effect of outliers and changes in sample size among plants of different heights. In our professional practice we have favored Model 5.2 since it easily allows the expansion of its specification to include more variables influencing fruit production (see Chapter XI). Models such as 5.2 can be integrated with models of other vital rates to evaluate population dynamics (see Chapter XIII). The reader may decide on other models using different criteria or applications.

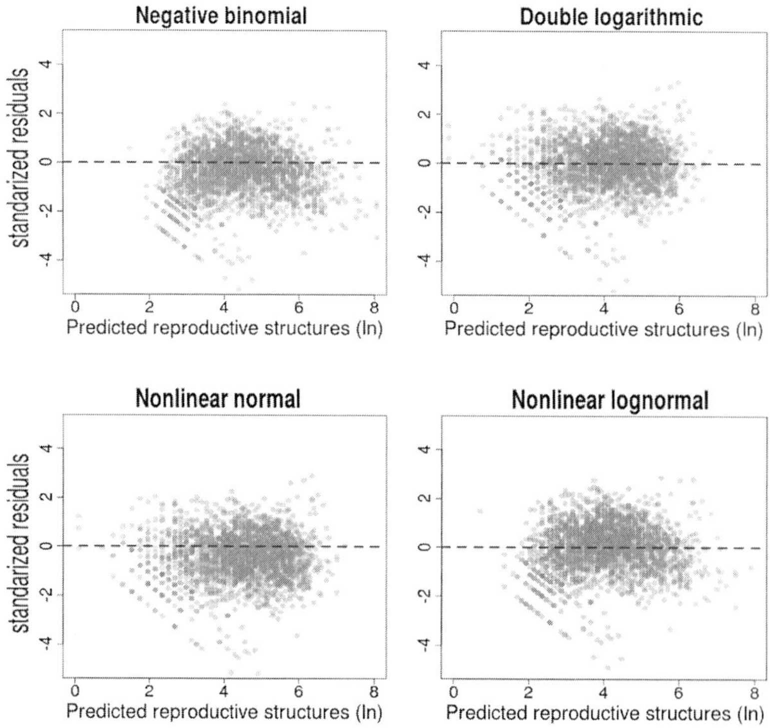

Figure 5.7: Standardized residuals of Models 5.2, 5.3, 5.4 and 5.5

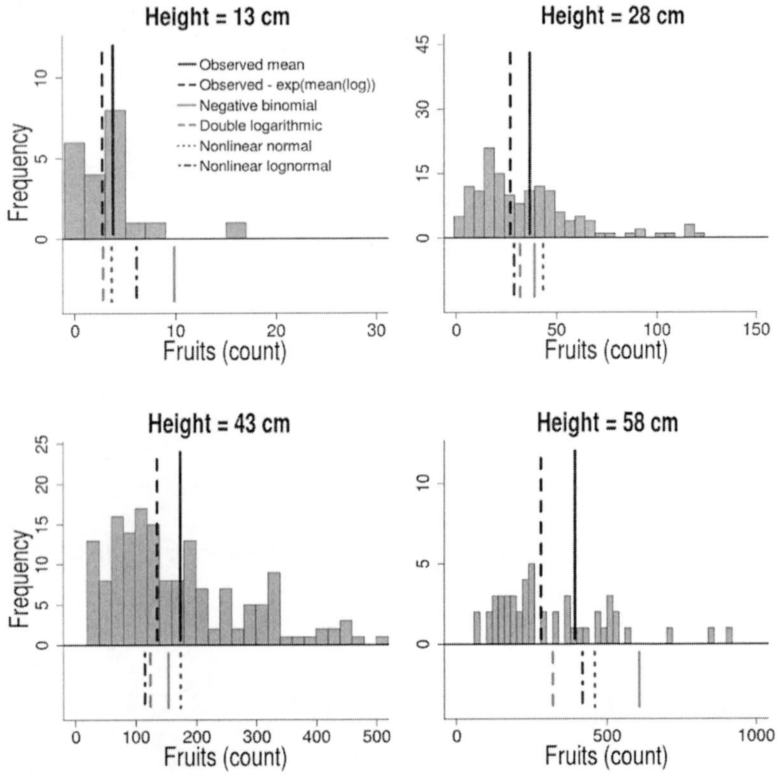

Figure 5.8: Histograms of the observed number of fruits as a function of heights of 12-14, 27-29, 42-44, and 57-59 cm. The lines below plots indicate the location of the posterior mean for Models 5.2, 5.3, 5.4 and 5.5 predicted for the height value at the center of the intervals (13, 28, 43 and 58 cm). The solid black line is the observed mean for the height interval and the dashed black line is the exponential of the observed mean of the same data logarithmically transformed

Independence of observations

There are more aspects of the model assumptions that we should inspect. These data represent several years of work in several populations, and our biological experience hints that there is a certain level of non-independence in the data because of variation in space and time. In Chapter IX we evaluate the effects by population using models considering 14 populations in different Florida scrub patches. To address the issue of non-independence among and within populations, we use models with a specification like that in Model 5.2 with the natural logarithm transformed data comparing responses in different scrub patches and evaluating random effects by population as deviations from the intercept and the slope.

Final considerations

Model assumptions have significant consequences on our inference and should be understood and evaluated to minimize misleading judgments. Here, we use the variation of fruits as a function of plant height of a rare Florida scrub species to demonstrate the consequences of different assumptions on model specifications. We exemplify serious problems of a suboptimal linear model and compare models with different likelihood distributions (including some that are non-compatible) for the data.

CHAPTER VI

THE USE OF PRIORS: PLANT HEIGHT AND THE NUMBER OF TILLERS IN WIRE GRASS

"If you wish to make an apple pie from scratch, you must first invent the Universe."
– Carl Sagan

Relevant notions

The use of priors is central and one of the greatest advantages of Bayesian approaches. Since their inception, scientific endeavors have generated knowledge by evaluating new ideas that build on previous findings. Bayesian priors provide one way to formalize this process, which constitutes the basis of scientific progress. There is an increasing appreciation that statistical inferences and predictions are best conducted using Bayesian methods (Dorazio 2016). However, insufficient instruction, computational limitations, misconceptions regarding the role of priors, and genuine concerns for the potential of misconduct have contributed to taper its widespread use. As the example from Chapter IV showed, priors improve our inference by combining previous relevant information with new data – this is particularly true when used to analyze small sample sizes.

There is an interesting debate to be had regarding the place of Bayesian inference within the philosophy of science. It might seem natural to link Bayesian approaches with inductive reasoning, contrasting it to the deductive reasoning associated with hypothesis falsifiability in frequentist approaches. Inductive reasoning deals, after all, with plausible guesses derived from incomplete evidence. Bayesian inference deals with degrees of belief represented by distributions. There are some authors, however, who believe that considering Bayesian approaches as merely inductive is a mistake (e.g., Gelman 1999). This disagreement is partly rooted in the use of priors. When selecting a prior distribution, we describe our previous knowledge using one of a handful of parametric families. These distributions are not subjective beliefs but objective hypotheses and require testing. According to these critics, therefore, the aim should not simply be to assign posterior distributions to models but to check how they fit the data. Beyond philosophical qualms, priors generalize findings and constitute a key part of robust approaches to validate information when confronted with new evidence.

Most alleged concerns of potential wrongdoing in the use of priors are similarly applicable to other statistical approaches, and its avoidance requires the same general measures for ethical behavior. The occasional forced withdrawal of papers from academic journals corroborates that the use of misleading information to prejudice inferences has occurred in the past and, regrettably, will continue in the future. A serious discussion on scientific misconduct is beyond the aim of this book, but we encourage our readers to better inform themselves on these matters to minimize their negative consequences (see, for example, LaFollette 2000 and Gross 2016).

In this chapter, we illustrate the use of priors through a case examining allometric relationships between two different size measures in a grass species: plant height and the number of tillers. We first propose a hypothesis that summarizes the rationality of our expectations, then identify an informative model specification and a reasonable likelihood for our data. We then proceed to build priors using data from previous studies of the same species collected in other areas and with other goals. These priors are combined with the proposed likelihood to generate the posteriors of our chosen informed Bayesian model. We compare these posteriors to those from a model with diffused priors where we ignore any prior information. Mirroring the approach in Chapter IV, we also compare the inference of these diffuse and informed models but with implementations based on samples of different sizes. Here, we do not have long-term data as a reference, so we compare the models against each other. We end the chapter with a discussion on the benefits gained with this approach.

Practical insights

Wiregrass (*Aristida beyrichiana*, formerly called *Aristida stricta*) is an endemic grass in the southeastern United States. It is a dominant component of the herbaceous layer in several pine communities. As it contributes a significant amount of fuel and is highly flammable, this grass plays a major role in the fire dynamics of these ecosystems. In 2013, graduate students at the University of Central Florida measured the height (in centimeters) and counted the tillers – segments that include a growing meristem – of hundreds of five-month-old wiregrass clumps that were seeded in pots and scheduled for transplanting into the field. The analysis presented here was used as a baseline in a restoration study, evaluating the effect of micro-habitat and ridge elevation on wiregrass growth, survival, and fecundity to better understand the process of successful reintroductions. In this chapter, we develop a Bayesian model of an allometric relationship for this grass species.

Allometry researches the constrained relationships among body parts, their scaling with size, and their changes during development (Shingleton 2010). The description of allometric relationships and the understanding of processes generating them are essential topics in ecology and evolution. Most scaling relationships become linear when plotted in log-log scales and can be expressed as linear equations. Not all the allometries need to be linear or linearizable with logarithms, though, and such an assumption must be tested. In this chapter, we will evaluate different prior specifications. The coefficients in these models describe differences in the growth of focal parts and the relationships between them and to the whole individual.

As wiregrass individuals increase in size, they might devote resources to increasing the height of a few tillers, increase the number of shorter tillers, or increase both proportionally. These growth patterns might change with the age of an individual or vary according to habitat characteristics. Here, we begin to explore the relationship between height and the number of tillers by exploring a functional equation that might describe it and then comparing these findings to prior data.

As always, we first need to visually explore our data to help identify potential distributions to represent them. For this purpose, we plot a histogram of each focal variable and overlay a potential distribution (Figure 6.1). Five months after germination, the grasses in the pots had an average height of 39.0 cm (with a standard error = 0.38). Overall, these heights were very close to being normally distributed. These plants also had about 7 tillers on average (with a standard error = 0.1). The number of tillers seemed to closely follow a Poisson distribution.

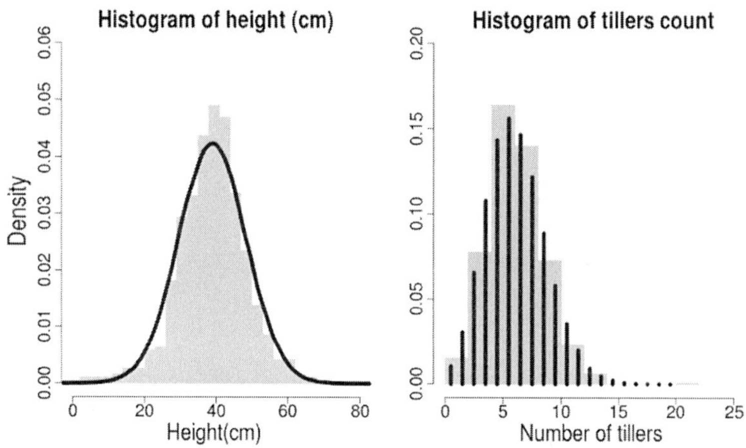

Figure 6.1: Left. Density plot (in gray) of observed plant height in cm and of a normal distribution (in black) with the same mean and variance. Right. Density plot (in gray) of the observed number of tillers and of a Poisson distribution (in black) with the same mean

Plotting grass height as a function of the number of tillers helps to convey a more complete picture of the overall data and the relationship between focal variables (Figure 6.2). There is a lot of additional information we can extract from these graphs. In the Cartesian plot (right) we can see the overall relationship between the number of tillers (in the x-axis) and plant height (in the y-axis), with the frequency distribution of each variable present as histograms in the margins. This plot indicates that the general relationship between our focal variables is non-linear and unimodal, but it also shows that there is a considerable spread of points across the plot for any value in either axis. Plants with fewer than five tillers tend to be small and plants with approximately 5 to 7 tillers reach the greatest heights, but as plants continue to increase in the number of tillers they decrease in height. The graph on the left is a tridimensional plot showing the frequency of points for every combination of the observed values for each variable. This plot emphasizes that the amount of information on this relationship is not consistent for all values across the ranges of the variables. Some combinations are less frequent (and therefore have more uncertainty) as we move away from the location of the average values for each focal variable. A good model specification should be able to account for this fact.

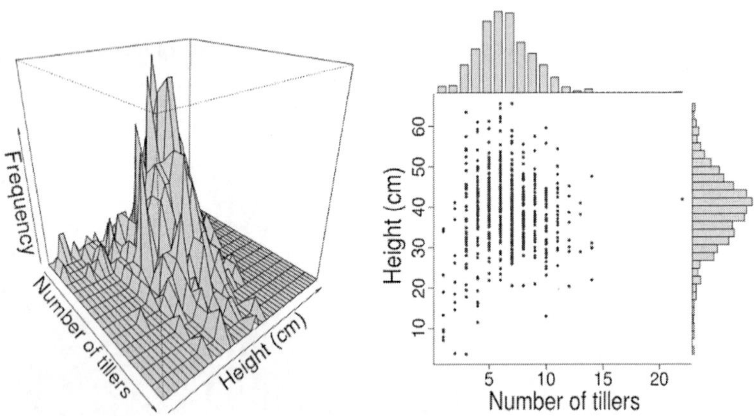

Figure 6.2: Plots of the number of tillers vs. plant height (in cm) for Wire grass

Perhaps the most striking feature of the tridimensional plot (left plot in Figure 6.2) is that it reminds us of a common assumption that is often accepted by researchers without further consideration. As discussed in Chapter IV, most statistical models assume that the predictor variable is measured without error. As we are dealing with allometric relationships, we have the option to use either variable as a predictor, but we could also evaluate a correlation model without this assumption. Smith (1973) developed a general specification to evaluate the relationship with error in both covariates. For the sake of convenience, we will embrace the assumption of no-error in the x-axis as it facilitates calculation and simplifies the model. Although we cannot guarantee that either variable was measured without error, we believe that counting less than 20 objects is less prone to error than measuring lengths, so we will use the number of tillers as the predictor variable.

Model building

We propose a quadratic polynomial function to model the relationship between our variables because of the observed shape of the data. Alternative specifications, which we do not explore in this chapter, include a linear relationship and an asymptotic exponential model. For the likelihood of these models, with and without informed priors, we use a normal distribution that is consistent with the distribution of the height data.

The chosen model specification evaluates a hypothesis on this relationship. We adhere to the ideas expressed by Betts et al. (2021) on the importance of stating and evaluating the support of our expectations to increase the chance of identifying underlying mechanisms and reduce bias. Reasonably, the model predictions implied by the quadratic polynomial model can derive from developmental, mechanical, and physiological constraints affecting the concurrent change in these variables. The proposed polynomial model has the form of a parabola. In this model, as the number of tillers increases, the mean height has an initial increase, followed by a maximum value, and then a decline. To make the intercept meaningful, we mean centered the number of tillers (i.e., we subtracted the mean from each observation). This displaces the mean to the position of zero in the x-axis making the intercept its prediction.

Our goal for this chapter is to demonstrate the process of building priors and show the advantages of using informed priors. For this purpose, we will evaluate two separate models: Model 6.1 (a simplified implementation is in Box 6.1) using diffuse priors, and Model 6.2, using informed priors. It is important to keep in mind that, regardless of the distribution used, we must define priors for every parameter estimated in a model. We evaluate these two models using all the available data (n=625) and compare the results to implementations with smaller sub-samples without the replacement of the same data (n=100, and n=40). Following the recommendations of Jenkins and Quintana-Ascencio (2020), we use samples larger than 25 to maximize the chances of identifying the most likely form of the relationships among these data.

The equation of the first quadratic polynomial with diffuse priors is:

$$height = a + b_1 \times tillers + b_2 \times tillers^2 \qquad Likelihood{\sim}N(\mu, \sigma)$$

Box 6.1: The model declaration in *Stan* for Model 6.1, including the specification of diffuse priors, is presented below (to see the entire code, check the annotated script on the book's website)

```
model{
    vector[N] mu;
    // sigma ~ uniform( 0 , 10 );
    b2 ~ normal( 0 , 10 );
    b1 ~ normal( 0 , 10 );
    a ~ normal( 0 , 100 );
    for ( i in 1:N ) {
        mu[i] = a + b1 * x[i] + b2 * x2[i];
    }
    height ~ normal( mu , sigma );}
```

Before we continue exploring the result of our first model, we should briefly discuss an important issue regarding priors. When defining their distribution, we must make sure we are not preventing the posterior from taking values that are plausible under our data. In our example, all the coefficients of the polynomial are given normal priors with a mean of zero and relatively large variances. The assumption here is that the posteriors will have means that fall within the range established by these prior distributions and that the variances of the posteriors will be much narrower. What constitutes a diffuse prior will vary from case to case. It might be easier to imagine potential problems in setting priors by considering our error coefficient (sigma). In the present example, we define the prior using a uniform distribution that gives equal probability to any value between 0 and 10 and a probability of zero to any value outside that range. If the range of the prior is too limited our prior definition would constrain the posterior to unrealistically small values. A good way to corroborate that our priors are behaving as expected is to plot the prior and posterior distributions side by side (see the end of the chapter).

Plotting the predicted average height as a function of the number of tillers with a 95% credibility interval shows that our first model does a good job at representing the overall trend of the data (Figure 6.3).

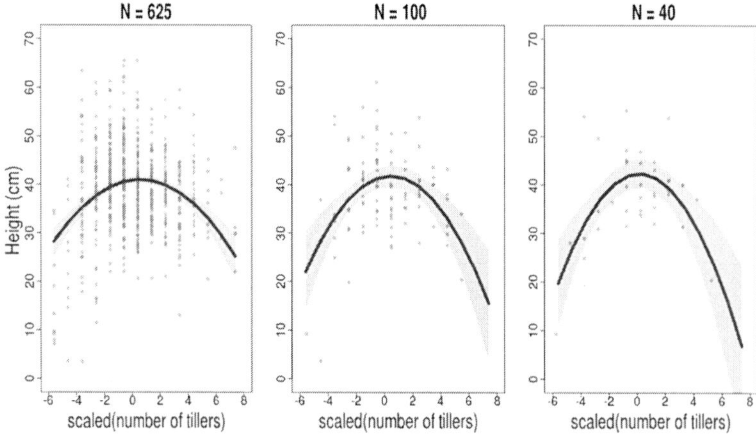

Figure 6.3: Plot of plant height (in cm) as a function of the number of tillers (centered) and Model 6.1 average predictions with its 95% credibility intervals for samples with a different amount of data (n= 625, 100, 40) Notice the location of the mean number of tillers (0 in the x-axis)

Informed priors

But what if we could get data to build informed priors? After all, we can't be the first people to ever work with wiregrass! Gordon and Rice (1998) collected similar data on the number of tillers and height of *Aristida beyrichiana* of seven different populations in Florida: Torreya State Park, Apalachicola Bluffs, Ravines Preserve, Wekiwa Springs (2 populations) and Rock Springs Run (2 populations). Researchers from the Jones Center in Georgia also collected similar data. They kindly allowed us to use their data (summarized in Table 6.1 and Figure 6.4).

Aristida beyrichiana is a long-lived species, and the data available in the literature covered a larger range of tiller numbers (up to 96 tillers) than our potted plants (maximum of 22 tillers). Curiously, some of the potted plants were taller (maximum height of 65.5 cm) than those in the literature (59 cm maximum height). This is likely due to the artificial growing conditions but might be important when interpreting results.

We could use these data in multiple ways to build our priors. We could use only those populations that were collected on sites close to ours and from a single cohort like ours. We could evaluate different model specifications and use the one that is more consistent among studies. Due to the relatively early age (only five months) of our potted plants, in this example, we use only data from the literature for plants with 22 or fewer tillers. In this way, we are more consistent with our measurement limit but keep our model specification and chosen likelihood. Readers may decide on different ways to use these data to build their models.

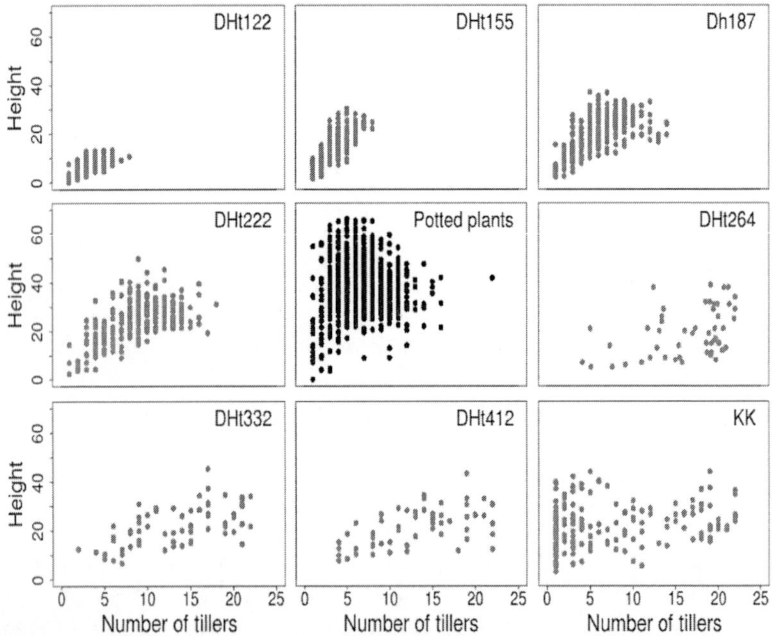

Figure 6.4: Plots of plant height (in cm) as a function of the number of tillers using data from the literature (Gordon and Rice 1998 and Kirkman, unpublished data) and our own data (center graph)

Table 6.1: Parameters for the quadratic polynomial model as specified in Model 6.1 calculated using our data and available data in the literature for height as a function of the number of tillers (variable x)

Population	**Intercept (a)**	**(b1) x**	**(b2) x^2**
Pop 1	9.94	3.54	-0.29
Pop 2	18.40	5.24	-0.29
Pop 3	33.49	5.68	-0.30
Pop 4	41.99	4.94	-0.21
Pop 5	43.22	2.88	-0.07
Pop 6	30.40	1.90	-0.04
Pop 7	34.52	2.51	-0.07
KK	22.57	0.28	-0.002
Mean	29.32	3.37	-0.16
Standard deviation	11.59	1.85	0.13
Our data	40.89	0.38	-0.34

🕶 Glasses On 🕶

We use the term diffuse priors (instead of noninformative priors) because every prior contains some level of information. They might simply indicate that any value in a large range is equally likely or provide some restrictions on the range of plausible values. Even when priors present more specific information about the values of a parameter, they can do so with different levels of uncertainty. Priors, therefore, can represent varying degrees of information. The more informative they are, the more they will influence the posterior.

The relative effect of the standard deviation for a diffuse prior approaches an asymptote as its value increases. It is possible to conduct a sensitivity analysis by running a model several times while varying the prior of interest. The goal is to check how changes in the prior affect the estimated posterior. In our first model, for instance, we can perform a sensitivity analysis for the prior of "b1" by testing different standard deviation values (Figure 6.5). Notice that for standard deviations above 1, the estimated posterior seems to reach an asymptote and remains constant. This is a good sign that our selected prior (with standard deviation = 10) is not having a marked influence on the posterior, and thus is effective as a diffuse prior. Some attributes of the focal sampled data and the model specification can also determine the influence of priors. The effect of priors on posterior distributions decreases as the number of the observed data and the likelihood of the parameters increase.

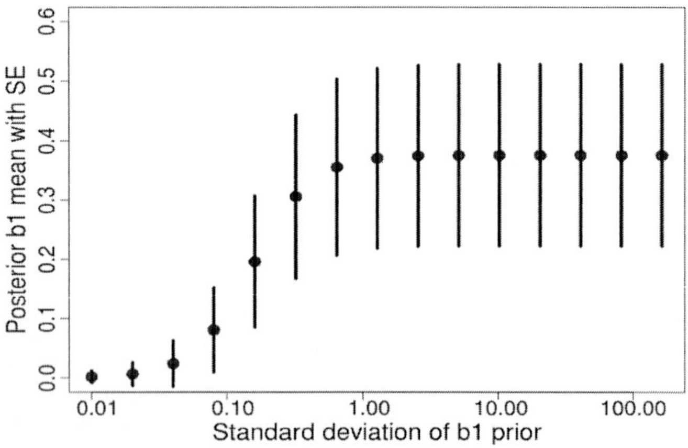

Figure 6.5: Change in posterior values of parameters b1 (and their standard error) as a function of the standard deviation value included in "diffuse" priors with mean zero

---

Because priors provide such a flexible way to incorporate previous knowledge, they allow for the use of diverse sources of information. Some potential sources include previous empirical studies (as used in this chapter; more examples in McCarthy and Masters 2005), results from meta-analyses, predictions from mechanistic models, and even expert opinion (see, for example, Choy et al. 2009). The process of gathering the information and formulating prior distributions varies according to the source. Luckily, there is extensive literature to guide you through this process.

We calculate the mean and standard errors of parameters generated with the same specification of Model 6.1 (Table 6.2) but using previously published datasets instead of our own and use the results to replace the specification of priors in the first model to build a new Model 6.2 with informed priors (Box 6.2).

Box 6.2: The model declaration in *Stan* for Model 6.2, including the specification of informative priors, is presented below (to see the entire code, check the annotated script on the book's website)

```
model{
     vector[N] mu;
     // sigma ~ uniform( 0 , 10 );
    b2 ~ normal( -0.16 , 0.13 );
    b1 ~ normal( 3.37 , 1.85 );
    a ~ normal( 29.32 , 11.59 );
    for ( i in 1:N ) {
        mu[i] = a + b1 * x[i] + b2 * x2[i];
    }
    height ~ normal( mu , sigma );
}
```

The coefficients obtained from our diffuse and informative models with different sample sizes (n=625, n=100, and n=40) are summarized in Table 6.2. Notice that the values for each coefficient have means within the suggested ranges and very narrow variances. Because we have mean centered the data, the value for parameter a (~40 across all models) represents the posterior mean prediction for the height of wiregrass with an observed average number of tillers (~6).

There are a few points worth noting from this table. First, the results from Model 6.1 and Model 6.2 are very similar when the entire data set (n=625) was used. These results are like our findings in Chapter IV where posterior predictions of models with and without priors converged as the sample size increased. Second, as sample sizes decrease, the use of informed priors also helps to maintain a lower level of uncertainty (lower standard deviations). Finally, and perhaps more importantly, the incorporation of informed priors better preserves the estimated mean of our parameters as the sample size decreases. When diffuse priors are used, the estimated mean of the polynomial starts drifting away from the values estimated with the full data set; for instance, b2 goes from -0.34 (when n=625) to -0.68 (when n=40). In contrast, when informed priors are used, the values remain close to those calculated with the full data set. In the case of b2, the estimated mean goes from -0.34 (when n=625) to -0.36 (when n=40).

Table 6.2: Estimated parameters of Models 6.1 and 6.2 and their uncertainty for different sample sizes

	n=625		n=100		n=40	
	Mean	**St dev**	**Mean**	**St dev**	**Mean**	**St dev**
Diffuse priors (6.1)						
a	40.9	0.43	41.69	1.07	42.42	1.55
b1	0.38	0.15	0.49	0.34	0.24	0.52
b2	-0.34	0.04	-0.55	0.12	-0.68	0.16
sigma	9.00	0.27	8.33	0.57	8.33	0.87
Informative priors (6.2)						
a	40.8	0.43	40.61	1.00	40.30	1.41
b1	0.39	0.15	0.50	0.35	0.65	0.50
b2	-0.32	0.04	-0.38	0.09	-0.36	0.11
sigma	9.01	0.25	8.43	0.59	7.80	0.87

These overall trends remain consistent when using other random subsamples (n=100 and n=40) of the same data. Here we only present one realization for demonstration purposes. We challenge the reader to evaluate other samples to verify this assertion. The posterior shape of the average predictions of all these models is consistent with a quadratic polynomial model. However, the use of the informed priors in Model 6.2 helps to maintain the overall shape of the response when sample size diminishes (Figure 6.6). In comparison, the response curve for Model 6.1 changed more radically with smaller sample sizes (Figure 6.3).

Notice in Figure 6.7, the displacement of the posterior of parameter *a* – the intercept – towards smaller values in Model 6.2 as the sample size decreases. This is consistent with the influence of the data used to build the prior. Remember that the plants in field conditions, as obtained from the literature, were on average smaller than our plants grown in pots. Figures 6.8 and 6.9 corroborate that the inclusion of the priors helps to preserve the posterior distribution of the coefficients involved with the shape of the quadratic polynomial. The difference is relatively minor for the b1 parameter, but the lack of informative priors displaces the posterior for b2, the second coefficient in the polynomial, to more negative values. The inclusion of an informative prior clearly reduced the uncertainty of this last parameter. Since most studies are based on relatively small samples, we are convinced that models with informed priors support greater generalizations and improved inference.

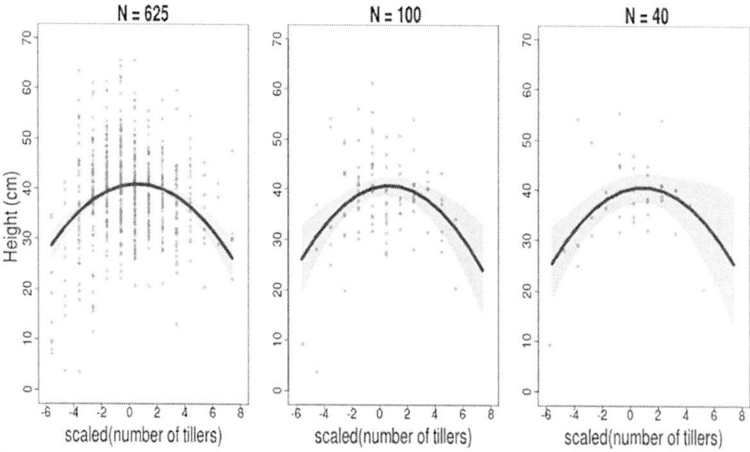

Figure 6.6: Plot of plant height (in cm) as a function of the number of tillers (centered) and Model 6.2 posteriors using informative prior predictions with its 95% credibility intervals

Final considerations

Both of our models support the notion of a constrained relationship between plant height and the number of tillers. We can think of mechanical, physiological, developmental, and ecological arguments that may help to explain our results. Perhaps there is a limited plant height which can mechanically support multiple tillers. It is also feasible that the cellular organization restricts the amount of tissue devoted to different functions. Selection pressure from competitors, herbivores and other environmental variables results in adaptive shapes maximizing fitness. Neither of our models can help to discern the mechanisms generating these results. However, the incorporation of informative priors in our second model helps to advance the notion that this might be a widespread phenomenon (and not just an artifact of our study).

As we mentioned earlier in the chapter, it is always important to corroborate that the priors are doing what we expect them to do. In the case of diffuse priors, we want to ensure that our choices don't have undue influence on the posterior. In the case of informative priors, we might want to get a sense of their relative contribution to the posterior. One way to do so – and one we recommend doing for all models – is to visualize the prior distribution particularly in contrast with the posterior (Figures 6.7-6.9).

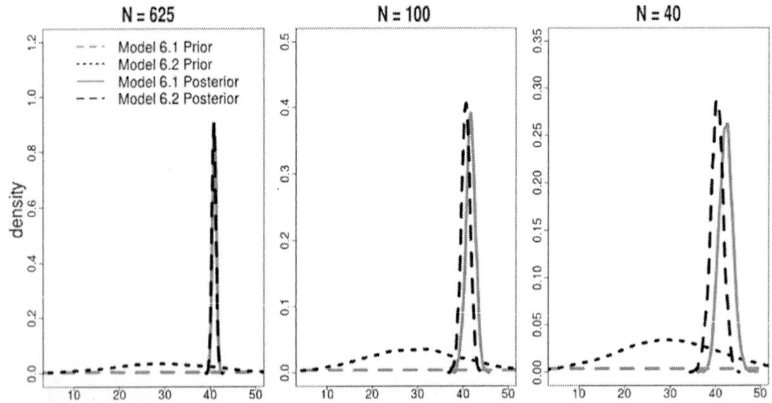

Figure 6.7: Distributions of the diffuse priors (grey dashed line) and informed priors (black dotted line), and the distributions of the respective posteriors, a gray continuous line for Model 6.1 and a black discontinuous line for Model 6.2, all for parameter a (the intercept)

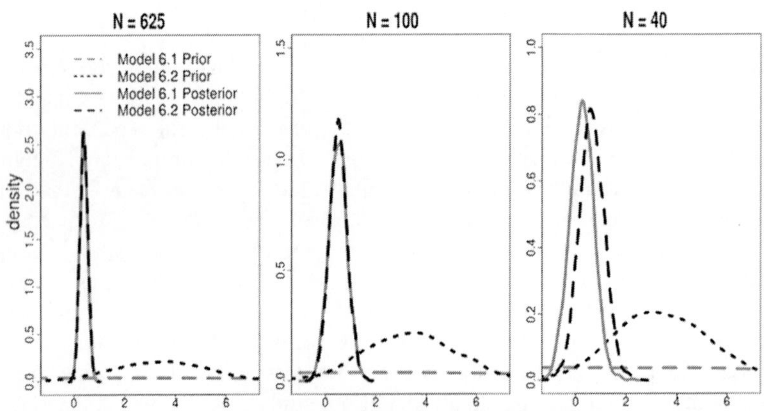

Figure 6.8: Distributions of the diffuse priors (grey dashed line) and informed priors (black dotted line), and the distributions of the respective posteriors, a gray continuous line for Model 6.1 and a black discontinuous line for Model 6.2, all for parameter b1 (the first coefficient of the quadratic polynomial)

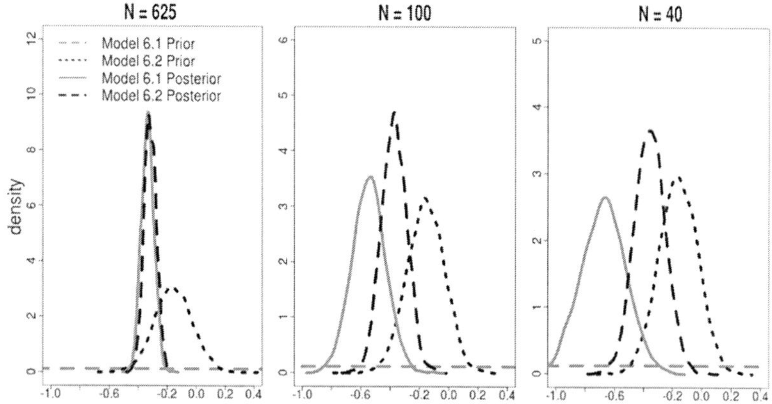

Figure 6.9: Distributions of the diffuse priors (grey dashed line) and informed priors (black dotted line), and the distributions of the respective posteriors, a gray continuous line for Model 6.1 and a black discontinuous line for Model 6.2, all for parameter b2 (the second coefficient of the quadratic polynomial)

Finally, we want to reiterate the importance of defining priors for every estimated parameter in a model. In the example presented in this chapter, both models required four parameters to be estimated: the 3 parameters of the polynomial describing the mean of the response (a, b1, and b2) and the error distribution of the response (sigma). If the available information allows us to build informative priors for only a portion of our models, we must still define diffuse priors for the rest.

CHAPTER VII

ANALYZING BINARY RESPONSES: TRADE-OFFS BETWEEN REPRODUCTION AND SURVIVAL

"... both the making and the unmaking were essential parts of life and necessary to keep the balance."
– Winona LaDuke

Relevant notions

Many phenomena manifest as binary options. Life or death, celibacy or reproduction, and the legendary outcomes of coin flipping are common examples. The data of these types of variables can be summarized as proportions and analyzed as probabilities. Many widely used methods for analyzing these proportions, however, have been less than optimal, particularly when trying to relate binary data to other variables.

An Ecology report by Warton and Hui (2011) found that 37% of 2008-2009 papers dealing with proportions used linear regressions on untransformed data. As you might suspect, this approach suffers from many drawbacks. For one, linear models produce unbounded predictions, which are inconsistent with the nature of proportions arising from binary variables. The fitted regression will very likely predict values below 0 or above 1 for some portions of the predictors. Another major problem is that the variance of a binomial proportion violates the assumption of homogeneity of variance required in linear models. One popular solution to address the issue of the variance has been to apply an arcsine transformation. The Warton and Hui (2011) report, found that this solution was used in 39% of papers dealing with proportions, making it the most used approach. For this solution to work, however, all proportions must be calculated using the same number of trials. If we have a numerical predictor, therefore, we must bin the data to ensure an equal number of observations per bin.

One big problem with both approaches discussed so far is the loss of information. When we transform counts into proportions, we lose track of sample size (5 out of 20 is the same proportion as 45 out 180). Luckily, a better way of analyzing the variation of binary outcomes was achieved by the development of the *logit* function. This elegant approach, developed by Joseph Berkson in 1944, maps the range (0, 1) onto the range (-∞, ∞), making it easier to analyze its response to other variables. The *logit* function is defined as the logarithm of the odds ratio:

$$logit(p) = ln\left(\frac{p}{1-p}\right) \qquad \text{(Equation 7.1)}$$

The relative ease of computation of the *logit* provided an alternative to the *probit* function, which was proposed by Charles Sanders Peirce at the end of the 19th century. The *probit* is still a viable solution to fit binomial models. In this chapter, however, we explore the logistic regression, which uses the *logit* function as a link between the predictors and the response. The use of a link function and the probability distribution assumed for the data are the main differences with a linear regression model. The process requires us to estimate the *logit* of p using some linear predictors:

$$logit(p) = \beta_0 + \beta_1 \times X_1 + \beta_2 \times X_2 + \cdots + \beta_n \times X_n \qquad \text{(Equation 7.2)}$$

Back transforming Equation 7.1, we can calculate p, so that:

$$p = \frac{e^{\beta\_0+\beta\_1\times X\_1+\beta\_2\times X\_2+\cdots+\beta\_n\times Xn}}{1+e^{\beta\_0+\beta\_1\times X\_1+\beta\_2\times X\_2+\cdots+\beta\_n\times X\_n}} \qquad \text{(Equation 7.3)}$$

We can now find the values of each coefficient ($\beta_0, \beta_1, \ldots, \beta_n$) to estimate the p that produces the binomial response that best explains our data.

🕶 Glasses On 🕶

Let's explore how the logit function is able to map a variable that ranges from 0 to 1 onto a variable that ranges from -∞ to ∞. If we let p be equal to 0.5 (the mid-point of the range), we can see that the logit is equal to 0.

$$logit(0.5) = ln\left(\frac{0.5}{1-0.5}\right) = ln(1) = 0 \qquad \text{(Equation 7.4)}$$

As p becomes smaller, its logit becomes negative. For instance, if $p = 0.2$, the logit will be -1.386. The closer p gets to 0, the more negative the logit becomes. On the other hand, when p takes a value greater than 0.5, the logit becomes positive. For instance, if $p = 0.8$, the logit will be +1.386. And, as p gets close to 1, the logit increases towards ∞. The shape of the relationship between the two is shown in Figure 7.1. One thing to note is that p can never be equal to 0 (as the logarithm of 0 is not defined) or 1 (as dividing by 0 is not defined).

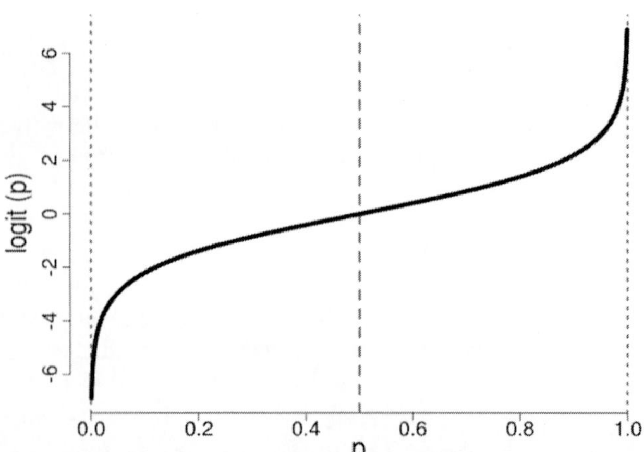

Figure 7.1: Graph of the logit function showing vertical asymptotes at p = 0 and p = 1 (dotted lines) and the inflection point at p = 0.5 (dashed line)

Practical insights

Ecological theory tells us that, given the restricted energy budget of organisms, there may be trade-offs between different vital rates (Stearns 1999). We often see trade-offs between survival and reproduction, which form the core of life-history strategies. An extreme example occurs in the Pacific salmon, which dies upon completing its reproduction. Salmon might stay at sea from 2 to 8 years before making the long journey upriver to reach their spawning grounds. Once they do, however, their probability of survival will go down to almost zero. The effort to complete the trip, which can be thousands of miles, and reproduce, can exhausts them to death.

Although the evidence for complete energy partitioning is controversial, individuals that commit to reproduction often grow costly structures for display and defense. To reproduce, they must engage in risky behaviors to find and preserve mates. When mating is successful, parents are frequently required to share their resources with the progeny to increase the chances that at least some will survive. Further, the presence of offspring can increase the chance of being attacked by predators. Generally, then, reproduction diverts energy that could otherwise be used to grow or increase the chances of the individual's survival. The proud parents are at higher risk relative to conspecifics that remain in celibacy, delay reproduction, or have fewer offspring.

Given that those who contribute more progeny to future generations are the ones that make a mark in phylogenetic trees, it is not surprising to see natural selection for multiple traits that increase fitness. We often see evidence of adaptations to delay reproduction until conditions are beneficial and adaptations to maximize the quantity or quality of progeny. The point is to minimize trade-offs, even if it sometimes involves imminent death for the parents. Different phenotypic attributes and habitat characteristics provide contrasting conditions for the expression and selection of these adaptations resulting in myriads of morphological and behavioral combinations (Judson 2003). Life history theory is one of the most important tenets in ecology and evolution with implications for human health, agriculture, resource management, and conservation.

In this chapter, we explore whether the reproductive effort of *H. cumulicola* individuals helps to predict the chance of surviving to the following season. In an environment where resources are scarce, expending too much energy on producing reproductive structures can have dire consequences.

We begin by plotting two histograms of the number of fruits: one for plants that survived to the following season and another for plants that didn't (Figure 7.2). We can readily notice that the histogram for surviving plants declines more steeply than the one for non-surviving plants. This might indicate an overall decline in the probability of survival with an increasing number of fruits.

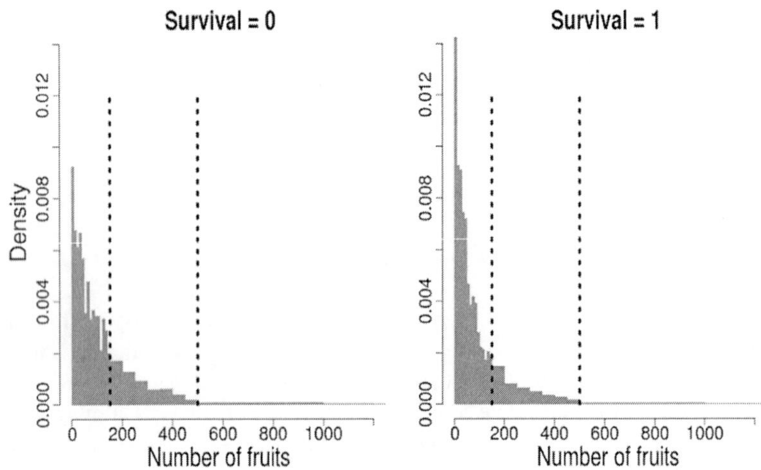

Figure 7.2: Left: Histogram of the number of fruits for plants that did not survive to the following season. Right: Histogram of the number of fruits for plants that did survive. The vertical dotted lines mark the limits for binned data groups in Figure 7.3

Model building

A logistic regression is a generalized linear model (GLM) with a binomial likelihood that can be used to analyze binary data such as survival (in the form of 0 and 1). All GLMs consist of the same three parts (Zuur et al. 2015): (1) a likelihood distribution, (2) a link function, and (3) a predictor function. The particular form of each element determines the type of regression and its role in modeling. For the logistic regression used in this chapter, these elements are defined as follows.

(1) The likelihood distribution is given by:

$$Survival_i \sim Bin(\pi_i)$$
$$expected\ (survival_i) = \pi_i$$
$$\text{variance}\ (survival_i) = \pi_i \times (1 - \pi_i)$$

(2) The link function for the logistic model is the logit of π. The estimate for the probability π of a variable as a function of X is given, in its simpler form, by:

$$\text{Logit}\ (\pi) = \eta$$

$$\pi_i = \frac{e^{\beta_0+\beta_1\times fruits}}{1+e^{\beta_0+\beta_1\times fruits}} \qquad \text{(Equation 7.5)}$$

(3) η is a predictor function of the covariates (X) and β_0 and β_1 are their model coefficients:

$$ln\left(\frac{\pi_i}{1-\pi_i}\right) = \eta = \beta_0+\beta_1 \times fruits \qquad \text{(Equation 7.6)}$$

The main purpose of this chapter is to demonstrate how to perform a logistic regression model in *R* and *Stan*. We use this approach to model the survival probability of *Hypericum cumulicola* with different numbers of fruits. We evaluate two alternative model specifications assuming different changes in survival as a function of the number of fruits. The dataset for this analysis (available on the book's website) contains the number of fruits of each sampled plant in a given year and a binary variable indicating the survival status of the plant at the following annual census (where 0 = dead and 1 = alive). Because the annual number of fruits was not counted for all sampled plants every year, we only use data between 1994 and 1997 when reproductive buds, flowers, and fruits were counted in every reproductive individual sampled. Since we want an estimate of the annual reproductive output of this plant, we combined these reproductive structures and used their sum as an approximation of fruits produced. In our demographic estimates, we multiply these numbers by an estimate of fruit set, but here we use the raw sums.

We build two Bayesian logistic regression models evaluating the effect of the number of fruits on the probability of survival of *Hypericum cumulicola* (Box 7.1). Model 7.1 assumes that the logarithm of the odds of survival changes as a linear function of the number of fruits. Model 7.2 assumes that the logarithm of the odds of survival changes as a power function of the number of fruits. While models with linear responses are emphasized in biometry courses, most biological phenomena have curve responses that are better represented with logarithmic data and non-linear models.

(Model 7.1)

$$\eta = \mathrm{a} + \mathrm{b} \times fruits \qquad likelihood\ Binom(p)$$

(Model 7.2)

$$\eta = \ln(\mathrm{a}') + \mathrm{b}' \times \ln(fruits) \qquad likelihood\ Binom(p)$$

Box 7.1: Constructing the code to run the analysis

We need to build two sets of code. The first one is an *R* script, including the following:

I. Defining the settings of the run

```
library(rstan) # it calls the library
rstan_options(auto_write = TRUE)
# it saves the compiled instance in our directory

orig_data <- read.csv("Hypericum_cumulicola.csv",
header=T)
options(mc.cores = parallel::detectCores())
# it runs the chains in parallel
dd  <- subset(orig_data, orig_data$fruits>0 &
time<1997)
dd$lnfruits <- log(dd$fruits)
```

II. Gathering the data for *Stan*

```
stan_data <- list(surv_fin=dd$surv_fin, rep= dd$rep,
N=nrow(dd) )
```

III. Setting the parameters monitored

```
params <- c("a","b")
```

IV. Defining the MCMC settings

```
ni <- 6000    # number of iterations
nt <- 2       # interval for data collection from the
MCMC chains
nb <- 2000    # number of iterations discarded
nc <- 3       # number of chains
```

V. Defining initial values

```
  inits <- lapply(1:nc, function(i) {
  list(a = rnorm(1,0,5), b = runif(1,0,5))})
```

VI. Calling for *Stan* from *R*
Model 7.1 should be replaced with Model 7.2 for the informed model

```
Model_7_1 <- stan("Chapter_VI_natural_scale.stan",
             data = stan_data, init = inits,
             pars = params,chains = nc, iter = ni,
             warmup = nb, thin = nt, seed = 1,
             open_progress = FALSE)
```

VII. Plotting and summarizing the posteriors

```
pairs(Model_7_1,pars=c("a","sigma"))
print(Model_7_1, digits = 3)
```

-- --

The second set of programs are two *Stan* scripts. They differ in a couple of lines. Below is the code for Model 7.1.

```
// Data declaration
data{
  int<lower=1> N;
  int survival[N];
  int fruits[N]; }
// Parameter declaration
parameters{
  real a;
  real b;
}
```

```
// Model declaration including prior specification
model{
  vector[N] p;
  b ~ normal( 0 , 10 );
  a ~ normal( 0 , 10 );
  for ( i in 1:N ) {
    p[i] = a + b * fruits[i];
  }
  survival ~ binomial_logit( 1 , p );
}
// Declaration of generated quantities
generated quantities{
  vector[N] p;
  real dev;
  dev = 0;
  for ( i in 1:N ) {
    p[i] = a + b * fruits[i];
  }
  dev = dev + (-2)*binomial_logit_lpmf( survival | 1 , p
);
}
```

-- --

In order to write Model 7.2, we replace the variable *fruits* with the variable *ln(fruits)* in the "Model declaration" and the "Declaration of generated quantities." Furthermore, we must not forget to change the variable that is specified as real numbers instead of integers (as logarithms are not whole numbers; Box 7.2).

Box 7.2: Data declaration for Model 7.2

```
data{
       int<lower=1> N;
       int survival[N};
       real lnfruits[N]; }
```

We provide diffuse priors and start values for both procedures. As always, we use the *R* function *rstan* to communicate with *Stan* software (the complete annotated code can be found on the book's website). We encourage you to inspect the trace plots of the MCMC chains for both models to corroborate that there were no problems in the generation of these estimates (see Chapter IV). Table 7.1 summarizes the estimates of both models:

Table 7.1: Average coefficients and their 95% credibility intervals for Models 7.1 and 7.2

Parameters	mean	Sd	Lower 2.5%	Higher 97.5%	Rhat
Model 7.1					
a	0.43	0.0536	0.32	0.53	1.00
b	-0.001	0.0002	-0.0014	-0.0005	1.00
Model 7.2					
a'	1.17	0.14	0.88	1.45	1.00
b'	-0.21	0.03	-0.27	-0.15	1.00

We use two criteria to evaluate the relative relevance of these models. First, we use observed proportions of surviving individuals binned at regular intervals of numbers of fruits per plant to assess how well our models predict these proportions (Figure 7.3).

We are not satisfied with the predicted posterior survival probabilities from Model 7.1 for many reasons. First, the curve fails to recover the marked decline in survival with an increasing number of fruits for plants with fewer than 50 fruits. The model also overestimates the survival of plants with 50-500 fruits. Finally, for plants with more than 700 fruits, the scarce information gives a few data points disproportionate leverage, pulling the survival curve down. The model predicts a probability of survival of approximately 0.3 for plants with around 1400 fruits. This is much lower than the observed annual proportion of surviving plants with 1400-1600 fruits in the previous census (~0.5; Figure 7.3).

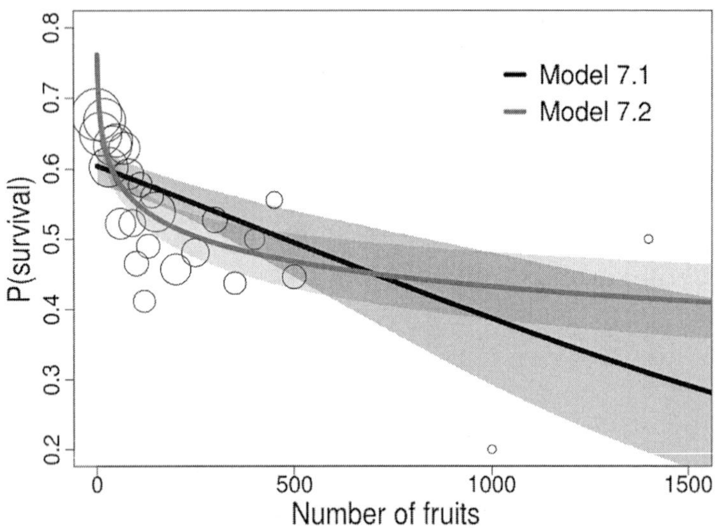

Figure 7.3: Survival probability as a function of the number of fruits. The circles are observed proportions for bins generated in three groups: between 1 and 150 (by 10), between 150 and 500 (by 50), and above 500 (by 200). Solid lines are average predictions for Model 7.1 (black) and Model 7.2 (gray). Grey polygons represent 95% credibility intervals for each model. The size of the circle is proportional to the amount of data in the bin

Model 7.2 is a power function, where the number of fruits has been log-transformed for analysis. It assumes that the change is proportional to the number of fruits and is fitted as a parabola (Figure 7.3). This model has a better match with the observed proportion of surviving individuals with a different number of fruits. It allows us to conclude that the trade-off between survival and reproduction is larger for plants with fewer fruits (most probably small plants) and becomes less steep with an increasing number of fruits.

The second criterion we use to evaluate the two models is to compare their WAIC (widely applicable information criterion) score. A generalized version of the better known Akaike information criterion, WAIC is better suited to estimate predictive performance in the context of Bayesian inference (Gelman et al. 2014). To calculate the WAIC score for each model, we extract the log-likelihoods from each model and use the *waic* function of the *loo* package (see the annotated script for more details; Vehtari et al. 2017). The results (Table 7.2) confirm that the second model is more likely (~23 times) to explain these data.

Table 7.2: Model comparison with WAIC scores for each model

	WAIC	dWAIC	weight
Model 7.2	2850.3	0.0	1.0
Model 7.1	2873.5	23.2	0.0

🕶 Glasses On 🕶

When comparing models, we are confronted with two types of statistical errors (McElreath 2016). **Overfitting** occurs when we interpret too much from the data by paying attention to peculiarities associated with the sample that are not generalizable to other samples. On the other hand, **underfitting** leads to poorer predictions than could be made from the information at hand. Measures of model fit – the ability of model predictions to match the sampled data – are not enough to resolve this issue. Increasing model complexity (by adding new parameters) improves model fit even if the extra variable represents a collection of random numbers. Yet, models that are more complex than warranted provide poorer predictions of new data than simpler models. Two possible approaches to deal with this conundrum are the use of **regularization priors** and **information criteria**.

Regularization is the use of weakly informed priors to improve inference and predictions by constraining parameters to reasonable ranges. Diffuse priors give the same plausibility to every parameter, allowing for posteriors that emphasize every detail of the data. Regularization priors help to reduce the risk of exaggerating predictive associations among variables (McElreath 2016). Used properly, then, regularizing priors can reduce overfitting by forcing the model to focus on regular features of the data. If regularizing priors are too skeptical, however, we might end up underfitting. A good strategy is to assess the strength of the data by testing several priors.

Information theory, which is used in this chapter, is centered on identifying ways to measure the reduction in uncertainty resulting from collected information. Model **deviance** is a measure of relative model fit and is defined as:

$$\mathrm{D} = -2 \sum_i \log\left(likelihood\ of\ case\ i\right) \qquad \text{(Equation 7.7)}$$

where likelihood is a mathematical expression that specifies the plausibility of the data. We can define two deviances for a model: one for the sample at hand (the test deviance) and another for the expectation of another random sample (the out-of-sample deviance). The Akaike information criterion (AIC) provides an estimate of the out-of-sample deviance (Burnham and Anderson 2002).

$$D_{out\ of\ sample}\,(AIC) = D_{test} + 2p \qquad \text{(Equation 7.8)}$$

where p is the number of parameters estimated in the model. The AIC is only one of many available information criteria. It is reliable when priors are flat, the posterior approximates a multivariate Gaussian, and the sample size is greater than the number of parameters. The AIC, however, was not developed as a Bayesian tool, and flat priors are rarely a good choice. The widely applicable information criterion (WAIC) is an alternative estimate of the out-of-sample deviance that is better suited for Bayesian models. It does not make assumptions on the shape of priors, does not require a multivariate Gaussian posterior, and provides analogous results to the AIC (Watanabe 2013). The WAIC is defined as:

$$WAIC = -2(lppd - p_{WAIC}) \qquad \text{(Equation 7.9)}$$

where *lppd* (log-pointwise predictive density) is the sum of the logarithm of the mean likelihood of each observation and pWAIC (effective number of parameters) is the sum of the variance of the log-likelihood for observation in the training sample. There are, however, some types of models, including Poisson and negative binomial models, for which the WAIC is not useful.

Final considerations

We presented evidence that increasing production of fruits reduces the chances of survival in *Hypericum cumulicola*. This is very much what was expected from ecological theory. Water and nutrient availabilities are temporary and spatially heterogeneous and can become very scarce in the Florida scrub, a system with sandy soils and a seasonal rain distribution. We can expect that growing and maintaining a large number of fruits are costly for many scrub specialists. *Hypericum cumulicola* stems die back every year in the fall, at the beginning of the dry season when the period of reproduction ends. New stems develop in the spring, with the beginning of the rainy season. Most of these new stems soon die, and the few remaining ones will support the reproductive structures that will disperse their seeds between June and September. This cycle also supports the occurrence of a trade-off between survival and fecundity in this species.

While both models showed the expected decline in survival probability with increased reproductive output, the power model was able to recover the non-linear pattern of this relationship. There are many reasons why biological phenomena are best modeled with non-linear models. An organism might be best suited for a particular range of conditions, but still subsist outside of that range. The response to an environmental factor might be contingent on another factor. There might be physiological or mechanical factors that impose an asymptote on the response. Whatever the reasons, many of these non-linear responses are best represented by logarithmic relationships. We recommend our readers to be willing to consider alternative models before committing to inference based on a few standard approaches. In our example, a model based on a power model function provides a better description of the relationship than the more standard form of the logistic regression using a linear predictor.

CHAPTER VIII

RANDOM EFFECTS OF POPULATIONS: REVISITING REPRODUCTIVE OUTPUTS

"A designer knows she has achieved perfection not when there is nothing left to add, but when there is nothing left to take away."
– *Antoine de Saint-Exupéry*

Relevant notions

In *Patterns of Culture*, the anthropologist Ruth Benedict famously expressed that each culture takes a few characteristics from "the great arc of human potentialities." Simply stated, even though cultures are unique, they all arise from a common element – the human experience – and are thus alike in some ways. If we travel abroad and come across a different society, we would not be completely naive about what we encounter. For example, we would not be baffled if we saw a group of people uttering modulated sounds from their mouths. We would immediately realize they are talking to each other, even if we didn't speak the language. Humans everywhere are bipedal animals who count their lifespans in decades and have common physical capacities and physiological needs. So, it would be reasonable to expect every society to function in accordance with those parameters. On our trip, we would find elements that are exotic to us but none that are otherworldly.

The notion that things are organized at multiple levels – being more similar to each other at lower levels while still being characterizable at higher levels – is the basis of mixed effect models. Phrases within a language, for instance, share syntactic and phonetic rules, but they also possess some universal properties such as using sounds that humans can make. As any polyglot can attest, learning multiple languages allows you to make some inference about all human languages while simultaneously deepening your understanding of any given one. This framework is appropriate and useful in most, if not all, ecological contexts.

In Chapter V, we evaluated models of reproductive output in a naive way, assuming that individuals were independent across populations and years. However, we recognized that plants in the same population are likely to be more like each other than, and consequently different from, individuals in other populations. Mixed effect models employ random effects to characterize the variation across a higher level (in this case across populations) from our sample.

Let's briefly discuss plants in different populations. Imagine we were interested in studying the effect of plant size on flowering. With the aim of generalizing our finding across the whole range of the species, we could count the number of fruits of individuals of a species and measure their height in different sites. Because populations differ in such things as population density, genetic makeup and demographic structures, neighbor composition and disturbance history, we could expect them to be different from each other. We could, therefore, use a model with dummy variables to handle location identity as independent categories, much like we have done in previous chapters. Such a model, however, has a fundamental problem, as it precludes us from drawing conclusions about the relationship between two or more variables (in this case the study of the effect of plant size on flowering) that can be extended to other populations of the species. In other words, the model assumes that parameters for a population are completely independent from those of others. Considering each population as completely autonomous of the others is a naive and poor way of learning about our surroundings. Pooling data would also be problematic since, similarly, we ignore valuable information at population level. Because random models characterize the variation among populations, they allow inference that is applicable to the conglomerate of populations.

Mixed-effect models offer a way to learn about individuals in each population while accounting for the pattern across populations. These models use data to estimate parameters describing the distribution of the response across populations and use that information to estimate the parameters within each population. Such an approach is extremely useful for answering management and conservation questions where often the focus is the responses across the whole distribution of a species.

-- --

🕶 Glasses On 🕶

Mixed-effect models recognize multiple sources of variability in the data. If data are clustered according to some criteria, mixed-effect models allow us to describe the effect of any common predictor variable (fixed effect) within each cluster as well as the variability among clusters (random effects). In our example, we might be interested in predicting flowering for individuals as a function of plant size. In this case, our fixed effect would be the plant size and our random effect the identity of the population where the plant occurs. The algebraic form of the simplest linear mixed-effect model in this case would be:

$$fruits_{ij} = \beta_1 + Pop_j + \beta_2 \times height_i + \varepsilon_{ij}$$

What differentiates this model from those seen in previous chapters is the addition of the random parameter (*Population* $_j$). Here, they are assumed to be random samples from a normal distribution and represent deviations from the intercept – each population is defined as having higher or lower estimates than the overall or reference mean depending on the model specification (set by β_1). Random effects, however, can also be used to modify slopes, so that the effect of the individual height might be different in each population (intercept: *Population* $_{1j}$ and slope: *Population* $_{2j}$).

$$fruits_{ij} = Pop_{1j} + \beta_1 + [Pop_{2j} + \beta_2] \times height_i + \varepsilon_{ij}$$

-- --

While evaluating ecological data, we are often solely interested in a few variables that relate to our goals. Concerned with obtaining information that is general and representative of our focal systems, we collect data in many locations and during numerous years. Robust designs will include enough replications to characterize the variation among these conditions.

When years and populations are considered in the study design to only have a reasonable representation of the ecosystem, then these variables are treated as random effects. Kéry and Schaub (2012) have an excellent discussion on the relevance of random variables and the different reasons we might have to use random effects in our models.

When we collect demographic data to describe the variation in vital rates of a species, we need to try to sample sufficient random sites and accumulate enough years to represent their spatial and temporal variation. Few locations and years will preclude a proper account of vital rate variation (Quintana-Ascencio et al. 2020). Following the available literature and our experience we recommend a minimum of 5-10 levels for these variables (Kéry and Schaub 2012).

In summary, random effects are parameters that we expect to belong to a common distribution. They are samples of a common distribution that arise due to some stochastic process. If we repeated our study, the sampled populations or years might change, but the underlying stochastic process described by their distributions would remain the same. The main idea behind treating a parameter as a random effect, then, is that we consider our groups (populations, years, etc.) to be just a sample of all possible groups. Further, we believe that the difference between groups arises from some random process, and that belonging to a group has a common impact on all observations within that group. Among many other benefits, random effects allow us to account for latent effects, avoid pseudo-replications, and increase the scope of our study beyond the groups of samples used (Kéry and Schaub 2012).

Practical insights

In Chapter V we demonstrated the importance of evaluating and understanding the consequences of model assumptions. We used data of *Hypericum cumulicola* to evaluate models and to predict the number of fruits of individuals as a function of individual height. We assumed that the individuals were independent across populations and years. However, we recognize that plants in the same population are likely to be more similar to each other than (and consequently different from) individuals in other populations.

The dataset and scripts for this example can be found on the book's website. The data for this example consist of the standing height in centimeters of each sampled plant and the number of fruits counted per individual. The dataset contains information about individuals in 14 populations of Florida rosemary scrub patches at Archbold Biological Station measured during August between the years 1994 and 1996. Here, we introduce a method to incorporate the population effects in our models. As there are only three years of data for this demonstration, we ignore the effects of years in this implementation. A model that evaluates random effects for years and populations and the effects of other variables on the fruit production can be found in Quintana-Ascencio et al. (2019). Here, we start by exploring a model that pools the data across the studied populations, then moving to a model assuming populations are independent from each other, and finally exploring two models with population as a random effect, one with random intercepts and another with random intercepts and slopes.

Model building

Pooled data

We can start by estimating the coefficients of a model that combines the data of individuals from all populations. In this model we assume that the effect of populations on individuals is invariant – the same as estimating a common intercept and slope. In other words, we assume plants in every population behave in the same way.

The underlying statistical Model 8.1 (implementation in Box 8.1) is:

$$log(fruits)_i = \beta_1 + \beta_2 * log(height)_i$$

$$Likelihood \sim N(\mu, \sigma) \qquad \text{(Model 8.1)}$$

Box 8.1: Simplified implementation in *R* and *Stan* for Model 8.1

```
model{
vector[N] mu;
// sigma ~ uniform( 0 , 10 );
b ~ normal( 0 , 10 );
a ~ normal( 0 , 100 );
```

```
    for ( i in 1:N ) {
        mu[i] = a + b * lgh[i];
    }
    ln(number of fruits) ~ normal( mu , sigma );
}
```

The results of Model 8.1 are presented in Table 8.1. The output of the general model of the pooled data should be familiar as we implemented it before in Chapter IV. This model explains approximately 66% of the variance. We plot the posterior means of this model in Figure 8.1.

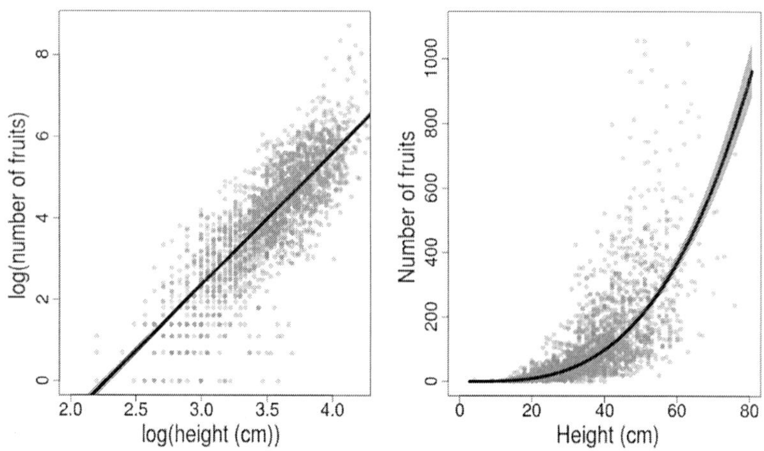

Figure 8.1: Left: Plot of ln (height) vs. ln (number of fruits) and predicted posterior means of Model 8.1 with pooled data (the mean as a black line and the 95% credibility interval as a grey polygon around this line). Right: The same model in natural scale (ordinate truncated to emphasize the bulk of the data)

Table 8.1: Results of Model 8.1 obtained from running three MCMC chains with 1000 sampling iterations

Coefficient	Mean	Std Dev	lower CI	upper CI	Rhat
a	-7.1	0.18	-7.4	-6.8	1.00
b	3.2	0.05	3.1	3.2	1.00
sigma	0.8	0.01	0.8	0.8	1.00

Population specific formulation

We now estimate the coefficients of the model per population. This is equivalent to doing a separate model for each population. Such an approach has several disadvantages. First, using specific intercepts (*Beta 1*) and slopes (*Beta 2*) for each population significantly reduces the overall degrees of freedom. Second, it greatly reduces our ability to draw inference beyond each population, limiting the scope of our interpretation.

When we plot the models for three example populations (Figure 8.2), we readily notice clear variation between each population. The coefficients per model by population are listed in Table 8.2. Notice that there is one population (57) with extreme deviations in the coefficients probably because of peculiar attributes and/or a smaller sample size. Observe below that in Model 8.2 the *k* indices of the coefficients are population specific.

$$log(Number\ of\ fruits)_i = \beta_{1_k} + \beta_{2_k} * log(height)_{ki}$$

$$Likelihood{\sim}N(\mu, \sigma) \qquad \text{(Model 8.2)}$$

Table 8.2: Coefficients estimated per population (popn)

Popn	**1**	**29**	**32**	**42**	**50**	**57**	**59**
Beta1	-9.8	-6.5	-11.3	-7.0	-10.9	-14.1	-7.0
Beta2	3.8	3.0	4.3	3.2	4.2	5.0	3.2
Sigma	0.7	0.7	0.8	0.6	0.8	1.3	0.7
N	91	141	179	169	189	94	174

Popn	**62**	**67**	**87**	**88**	**91**	**93**	**103**
Beta1	-6.9	-6.7	-7.8	-9.7	-5.5	-6.8	-6.0
Beta2	3.1	3.0	3.3	4.2	2.7	3.0	2.8
Sigma	0.6	0.6	0.8	0.7	0.8	0.8	0.8
N	135	122	129	140	183	206	149

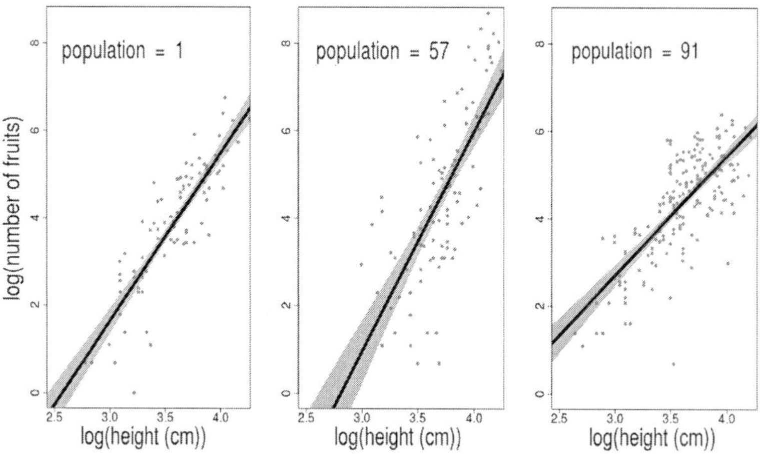

Figure 8.2: Plots of ln (height) vs. ln (number of fruits) and predicted means of iterations of Model 8.2, for example populations (populations: 1, 57 and 91 respectively; the mean as a black line and the 95% credibility interval as a grey polygon around each line). Data as points in grey. See Table 8.2 for a complete list of populations

Mixed models with random variation by population

As we are interested in drawing more general inferences (and making better use of the data), we can instead estimate the variation around the coefficients (the intercept or both the intercept and slope) due to random variation across populations. Here, we assume that the deviations of the effects by population are normally distributed. In this way, we will inform the estimate of one population with that from other populations. For more details on these assumptions, see Zuur et al. (2009). We start by assuming random intercepts but a common slope across population.

$$log(fruits)_i = \beta_1 + \alpha_{1k} + \beta_2 * log(height)_{ki}$$

$$\alpha_{1k} \sim N(0, \sigma_1) ;$$

$$Likelihood \sim N(\mu, \sigma) \qquad \text{(Model 8.3)}$$

A visual exploration of Model 8.3 (Figure 8.3) helps us to better understand the implications of modeling random effects only for the intercepts.

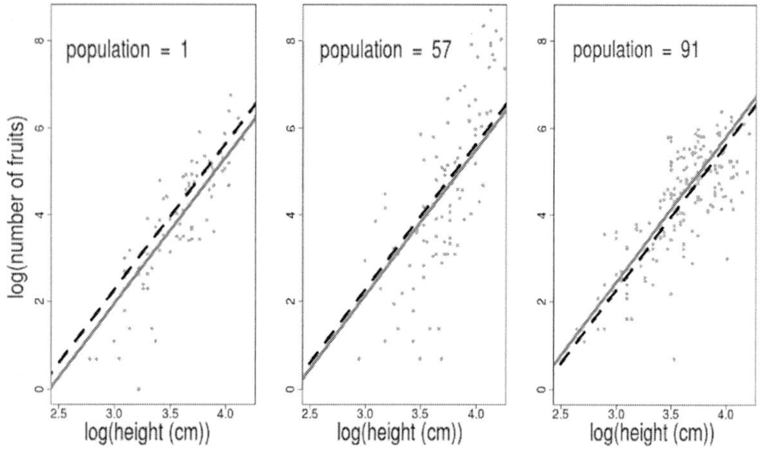

Figure 8.3: Plots of ln (height) vs. ln (number of fruits). The predicted overall mean of Model 8.3 is shown as a dashed black line. The predicted mean for the selected population is shown as a dark grey line

The dashed black lines (which are the same in all 3 panels) are the mean posteriors obtained only with the fixed coefficients of Model 8.3 (the mean population effect). The solid gray lines represent the variation estimated by population as their displacement from the average curve. The random intercept models are curves that shift vertically by a factor that is normally distributed with a given variance. If the deviation from the average is larger, the shift is greater. This shift, in this case, is the same for plants of all sizes.

Table 8.3: Mean coefficients estimated by population (mean [Beta1] + deviation by population) for mixed Model 8.3 with variable intercepts. Mean fixed coefficients: Beta1 = -7.48, Beta2 =3.29, and sigma intercept = 0.29, sigma = 0.78

Population	**1**	**29**	**32**	**42**	**50**	**57**	**59**
Beta1	-8.1	-7.9	-7.1	-7.8	-8.0	-7.9	-7.7

Population	**62**	**67**	**87**	**88**	**91**	**93**	**103**
Beta1	-7.9	-7.7	-7.8	-7.9	-7.6	-7.6	-8.0

Finally, we can explore a model that estimates random intercepts and slopes by population, Model 7.4. This is specified in the following equations.

$$log(fruits)_i = \beta_1 + \alpha_{1_k} + [\beta_2 + \partial_{2_k}] * log(height)_{ki}$$

$$\alpha_{1k} \sim N(0, \sigma_1)$$

$$\partial_{2k} \sim N(0, \sigma_2) ;$$

$$Likelihood \sim N(\mu, \sigma) \qquad \text{(Model 8.4)}$$

A visual exploration of Model 8.4 (Figure 8.4) reveals the differences from the random intercept model (Model 8.3). Once again, the dashed black line represents the model obtained with the fixed (average) coefficients. The lines in gray represent models in which both the intercepts and the slopes of each population are displaced from the values of the average curve by factors that are normally distributed. Notice that this model does not evaluate other possible effects such as the possible spatial correlation between populations. We will explore this issue in Chapter XI.

Comparing all the models that contain fixed effect parameters for the combined data (Table 8.4), we can see that Model 8.4 with random intercepts and slopes has the lowest WAIC score, indicating that it is the most informative of the set.

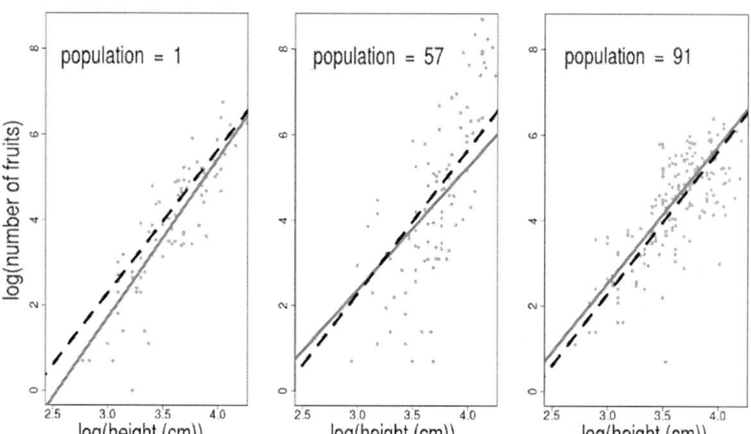

Figure 8.4: Plots of ln (height) vs. ln (number of fruits). The predicted overall mean of Model 8.3 is shown as a dashed black line. The predicted mean for the selected population is shown as a dark gray line

Model comparison

Model 8.4, the random slope and intercept model, avoided losing information and identified variation across populations. It allowed us to better characterize the overall variation of the change in the number of fruits with plant height (Figure 8.5). Simultaneously, it permitted us to estimate this variation for each population with the advantage that we improve the estimates of populations with low sample sizes by "borrowing" information (McElreath 2016) from other populations – similar to a polyglot who needs fewer lessons to learn a new language.

Table 8.4: Coefficients estimated for mixed Model 8.4 with variable intercepts and slopes. Fixed coefficients: Beta1 = -8.17, Beta2 = 3.46, sigma intercept = 2.05, sigma slope = 0.60, and sigma = 0.76

Population	**1**	**29**	**32**	**42**	**50**	**57**	**59**
Beta1	-9.5	-7.9	-9.4	-5.9	-6.9	-6.3	-6.8
Beta2	3.7	3.4	4.1	2.8	3.1	2.9	3.1

Population	**62**	**67**	**87**	**88**	**91**	**93**	**103**
Beta1	-10.5	-7.2	-10.4	-12.3	7.2	-7.1	-7.0
Beta2	4.0	3.2	4.1	3.5	3.2	3.2	3.1

Table 8.5: Comparison of the average coefficients of the population level for the three approaches (only fixed effects for mixed models), and their WAICs. The model with a random intercept and random slope (Model 8.4) was the most likely of the set

Models	**Intercept**	**s.e**	**Slope**	**s.e**	**WAIC**
pooled data	-7.1	0.2	3.2	0.1	5118
Random					
slope	-7.3	0.2	3.3	0.05	4953
intercept & slope	-8.1	0.6	3.5	0.2	4845

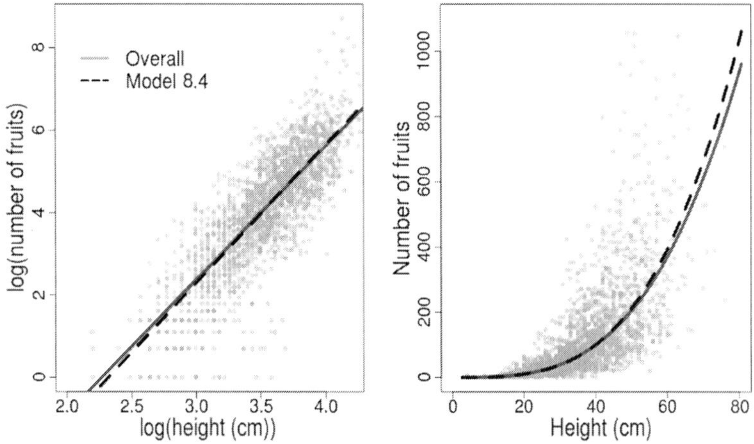

Figure 8.5: Left: Plot of ln (height) vs. ln (number of fruits) and predicted overall mean of Model 8.1 with pooled data (solid gray line) and predicted mean using fixed effects from Model 8.4 (dashed black line). Right: The same models in natural scale (ordinate truncated to emphasize the bulk of the data)

Final considerations

As we have seen in this chapter, mixed-effect models are those that incorporate both fixed effects and random effects making them very powerful to deal with ecological data which often include important temporal and spatial components. In other words, they allow us to learn about the overall pattern of interest (fixed effects) while accounting for and characterizing differences between subsets of data that can be interpreted as arising from a common process (random effects). Defining a variable as a random effect has several advantages, such as expanding the reach of model inference, improving estimates of data-poor groups, avoiding pseudo-replication and better accounting of correlations and latent effects.

Here, we evaluated the models using the logarithm of the data and assuming a Gaussian likelihood. We follow the tradition of scholars involved in allometric relationships who have demonstrated the convenience of these scales to describe rates between body measurements. However, since the numbers of fruit are counts, other model specifications with Poisson or negative binomial likelihoods are worth exploring (see Chapter V).

Despite the many advantages, there are circumstances when treating a variable as a random effect is not possible or appropriate. For instance, if we use a variable with very few levels, calculating a distribution would be very difficult. In other cases, the assumption that groups vary due to a common process (i.e., they are random draws from a common distribution) might be unrealistic. If two or more different groups exist in the data, with different characteristics, the use of random effects might mask any real differences; for example, if locations represent two different habitats such as roadside and native vegetation (see Chapter XIII).

Overall, the decision of whether to treat a variable as a random effect hinges mainly on three issues: (1) Any variable can be treated both as fixed or random depending on the research question. For example, a researcher might be interested in specifically quantifying the fixed effects of a certain categorical predictor on the response variable, regardless that such a variable could be potentially treated as a random factor, (2) the assumption that different levels of the variable represent random samples of a larger set, and (3) the number of levels in the sample is sufficient to characterize the parameters of the assumed distribution (most frequently the variance of a normal distribution with mean zero).

CHAPTER IX

COUNT DATA: SEEDS PER FRUIT

"Why so much effort in counting?
To avoid somebody deceiving me."
– *Francisco Villa (Doroteo Arango)*

Relevant notions

The outcome of many biological and ecological processes can be summarized as counts. These types of data are made up of integers bound to a lower limit, since no count can be less than zero. Counts are non-continuous variables whose variance frequently increases with the mean (Crawley 2007). This makes intuitive sense. If you have ever grown fruit trees, you have surely noticed that young trees produce far fewer fruits than fully mature ones. You probably also noticed much greater variation among the mature trees. Dwarf apple trees, for example, usually produce only a few dozen fruits in their first reproductive year, ranging from less than 10 to more than 100. A mature tree, in contrast, can produce over 500 apples on average, but with a much wider range (from a few dozen to several thousand).

Linear regressions are ill-equipped to deal with the bounded nature of counts and the lack of homogeneous variance. Generalized linear models (GLMs) once again offer a much better alternative to help evaluate a hypothesis for these types of responses. As it was introduced in Chapter VII, a generalized linear model is defined by three elements: the linear predictor, the link function, and the error structure. The estimated values are obtained with a transformation of the data and calculated with a linear predictor. The link function relates the values of the response variable to the linear predictor. These models allow us to specify different variance distributions to describe counts.

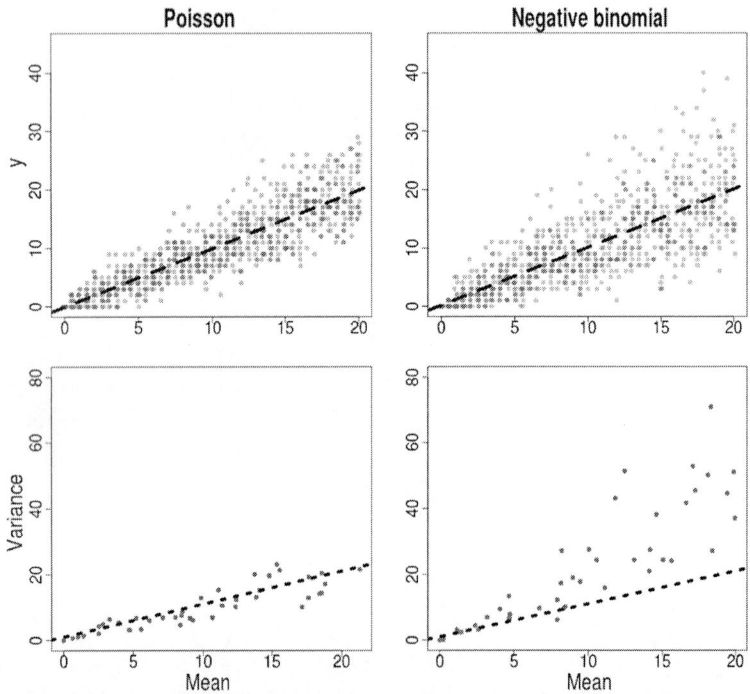

Figure 9.1: Examples of data described using a Poisson distribution (left) vs. a negative binomial distribution (right). Data as gray points on the upper plots. The dashed line is the mean value of y for each value of x. Variance on y for each point on x as gray points in lower plots. The dotted line is the identity line

One of the most common and useful probability distributions used for counts is the Poisson. It estimates the probability of obtaining a count x when the mean count per unit is λ (Crawley 2007), and it works fine when the mean is fairly equal to the variance (Figure 9.1, left plots). This, however, is not always the case. When the variance in counts is greater than the mean, the data are better described by the negative binomial distribution (Crawley 2007; Figure 9.1, right plots). The link function for these types of models is the logarithm. Notice in Figure 9.1 how the response y for the negative binomial fans out faster from the mean as the independent variable x increases (top plots). Notice also how, when comparing the mean and variance (bottom plots), the data stay close to the identity line for the Poisson but deviate upwards for the negative binomial.

Practical insights

Seeds are embryonic plants confined inside of a covering coat usually with some stored sustenance (Hanson 2015). Depending on the species, a single fruit may contain anywhere between one seed to thousands of seeds. Seed counts can help us to assess the magnitude of an individual's reproductive effort and, thus, are common chores among plant ecologists – particularly demographers. In this chapter, we will compare the seed number by fruit among plants of a herbaceous species in different sites and years.

Researchers at Archbold Biological Station collected seeds of *Hypericum cumulicola* in 12 different populations in August 1995, 1996, 1998, and more recently 2014. Here, we focus only on fruits that developed seeds. We are aware that the number of seeds by fruit is a relatively stable trait among plants of the same species (Hanson 2015), but we still speculate that resource variation among populations and years may affect the expression of this attribute. We, therefore, consider these spatial and temporal effects as random and assume them to be deviations from the estimated overall mean (see Chapter VIII). Considering all the data, the aggregate counts from 989 fruits have a mean of approximately 13 seeds (Figure 9.2). The observed variation is different from those obtained with the most likely Poisson and negative binomial distributions for these data (not including any covariates). These extra deviations may be associated with variation due to years and population, besides other factors such as count error. We are interested in assessing the structure and magnitude of these deviations.

As individual populations and years are interchangeable elements in this design, we assume them to be random effects. Each group is assumed to be sampled from a normal distribution. The distributions of the deviations from the mean for these two factors have a mean of zero, and we estimate the common variance for each group: years and populations.

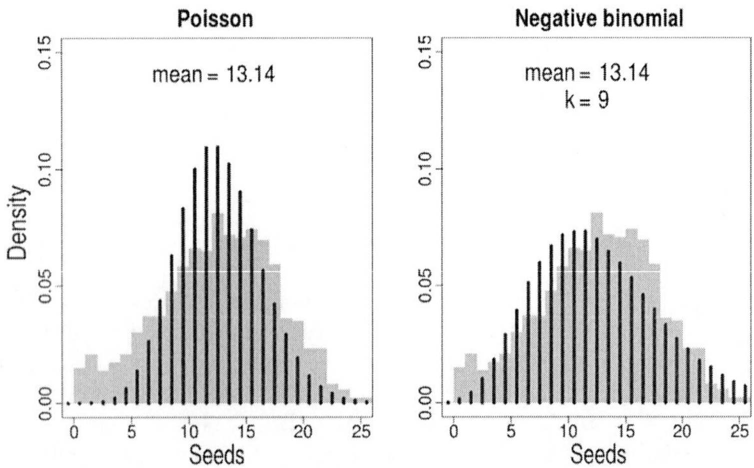

Figure 9.2: Histogram of the number of seeds by fruit in *H. cumulicola* (data pooled across years and populations; n=989 fruits). In gray is the density of the observed counts, in black is the density of a Poisson (left) and a negative binomial distribution (right) with the most likely parameters

Model building

Assuming a Poisson distribution for the likelihood of these data and given a model that contemplates our question, the elements of a generalized model are:

$$N\ seeds \sim Poisson(\mu_{ij})$$

$$E(N\ seeds) = Var(N\ seeds) = \mu_{ij}$$

Where μ_{ij} is the expected mean in population i and year j. The mean is equal to the variance. The link function is the logarithm of μ_i

$$\ln(\mu_{ij}) = \eta_{ij} \qquad \text{(Equation 9.1)}$$

The predictor η_i is a function of the covariates, in this case the populations and years:

$$\eta_{ij} = \beta_0 + \alpha_i[population_i] + \rho_j[year_j] \quad \text{(Equation 9.2)}$$
$$\alpha_i \sim N(0, \sigma_i)$$
$$\rho_j \sim N(0, \sigma_j)$$

where the population and the year are assumed random effects.

Assuming a negative binomial distribution for the likelihood of the data and applying the same model as above, the elements of a generalized linear model are given by:

$$N\ seeds \sim NB(\mu_{ij}, k)$$

$$E(N\ seeds) = \mu_{ij}$$

$$var(N\ seeds) = \mu_{ij} + \mu_{ij}^2/k \quad \text{or} \quad var(N\ seeds) = \mu_{ij} + \alpha \times \mu_{ij}^2$$

The link function is again the logarithm of μ_{ij}:

$$\log(\mu_{ij}) = \eta_{ij} \quad \text{(Equation 9.3)}$$

The predictor η_i is again a function of the covariates – populations and years.

$$\eta_{ij} = \beta_0 + \alpha_i[population_i] + \rho_j[year_j] \quad \text{(Equation 9.4)}$$
$$\alpha_i \sim N(0, \sigma_i)$$
$$\rho_j \sim N(0, \sigma_j)$$

And where the population and the year are assumed as random effects.

In Box 9.1 we present some of the key elements of the code in *R* used for these two models (for the full code, see the annotated script on the book's website). In particular, we want to focus on the parameters *xsqp* and *xsq* being generated for the Poisson and the negative binomial respectively. These are important for calculating the dispersion statistics (*dispersal_p* and *dispersal_b*) used to select between the two models. In Boxes 9.2. and 9.3 we present the full *Stan* code for both models (notice *xsqo* and *xsq* are generated within the model).

Box 9.1: Some highlights of the *R* code are:

II. Setting the parameters monitored

```
# for Poisson
params <-c("b0","p","y","sigmap","sigmay","xsqp",
"lambda")

# for negative binomial
params <- c("b0","p","y","sigmap","sigmay","xsq",
"scale","pbar")
```

#VI. Retrieving the output from *Stan* and calculating the dispersion statistic

```
# for Poisson
print(Model_8_1 ,digits=3,depth=2)

post1 <- extract.samples(Model_8_1)
dispersal_p <- post1$xsqp/(nrow(cd)-3)
plot(density(dispersal_p))

# for Negative binomial
print(Model_8_2 ,digits=3,depth=2)

post2 <- extract.samples(Model_8_2)
dispersal_b <- post1$xsq/(nrow(cd)-4)
plot(density(dispersal_b))
```

Box 9.2: *Stan* code for the Poisson model

// Data declaration

```
data{
    int<lower=1> N;
    int<lower=1> N_population;
    int<lower=1> N_yrs;
    int seeds[N];
    int population[N];
    int yrs[N];
}
```

// Parameter declaration

```
parameters{
    real b0;
```

```
    vector[N_population] p;
    vector[N_yrs] y;
    real<lower=0> sigmap;
    real<lower=0> sigmay;
}
```

// Model declaration including prior specification

```
model{
    vector[N] lambda;
    sigmay ~ cauchy( 0 , 1 );
    sigmap ~ cauchy( 0 , 1 );
    y ~ normal( 0 , sigmay );
    p ~ normal( 0 , sigmap );
    b0 ~ normal( 0 , 10 );
    for ( i in 1:N ) {
        lambda[i] = b0 + p[population[i]] + y[yrs[i]];
    }
    seeds ~ poisson_log( lambda );
}
```

// Declaration of generated quantities

```
generated quantities{
  vector[N] lambda;
  vector[N] pseeds;
  real dev;
  real xsqp;
  dev = 0;
  xsqp = 0;
  for ( i in 1:N ) {
    lambda[i] = b0 + p[population[i]] + y[yrs[i]];
    pseeds[i] = exp(lambda[i]);
  }
  dev = dev + (-2)*poisson_log_lpmf( seeds | lambda );
  for ( i in 1:N ) {
    xsqp = xsqp + pow(seeds[i]-mean(pseeds),
          2)/mean(pseeds);
    }
}
```

Box 9.3: *Stan* code for the negative binomial model

// Data declaration

```
data{
    int<lower=1> N;
    int<lower=1> N_population;
    int<lower=1> N_yrs;
```

```
    int seeds[N];
    int population[N];
    int yrs[N];
}
```

// Parameter declaration

```
parameters{
    real b0;
    real<lower=0> scale;
    vector[N_population] p;
    vector[N_yrs] y;
    real<lower=0> sigmap;
    real<lower=0> sigmay;
}
```

// Model declaration including prior specification

```
model{
    vector[N] pbar;
    scale ~ cauchy( 0 , 10 );
    sigmay ~ cauchy( 0 , 1 );
    sigmap ~ cauchy( 0 , 1 );
    y ~ normal( 0 , sigmay );
    p ~ normal( 0 , sigmap );
    b0 ~ normal( 0 , 100 );
    for ( i in 1:N ) {
        pbar[i] = b0 + p[population[i]] + y[yrs[i]];
        pbar[i] = exp(pbar[i]);
    }
    seeds ~ neg_binomial_2( pbar , scale );
}
```

// Declaration of generated quantities

```
generated quantities{
  vector[N] pbar;
  real dev;
  real xsq;
  dev = 0;
  xsq = 0;
  for ( i in 1:N ) {
      pbar[i] = b0 + p[population[i]] + y[yrs[i]];
      pbar[i] = exp(pbar[i]);
      xsq = xsq + pow(seeds[i]-exp(b0),
      2)/(exp(b0)+(pow(exp(b0), 2)/scale));
     }
  dev = dev + (-2)*neg_binomial_2_lpmf( seeds | pbar ,
scale );
}
```

We use different implementations for these models. The Poisson requires 3 parameters and their associated priors (one per parameter): the overall mean (*b0*); the variance of the random effects by population (*sigmap*); and the variance of the random effects by year (*sigmay*). The negative binomial model requires the same parameters plus one more for the scale (*k*) of the negative binomial equation. The posterior estimates for these parameters for both models are presented in Figure 9.3.

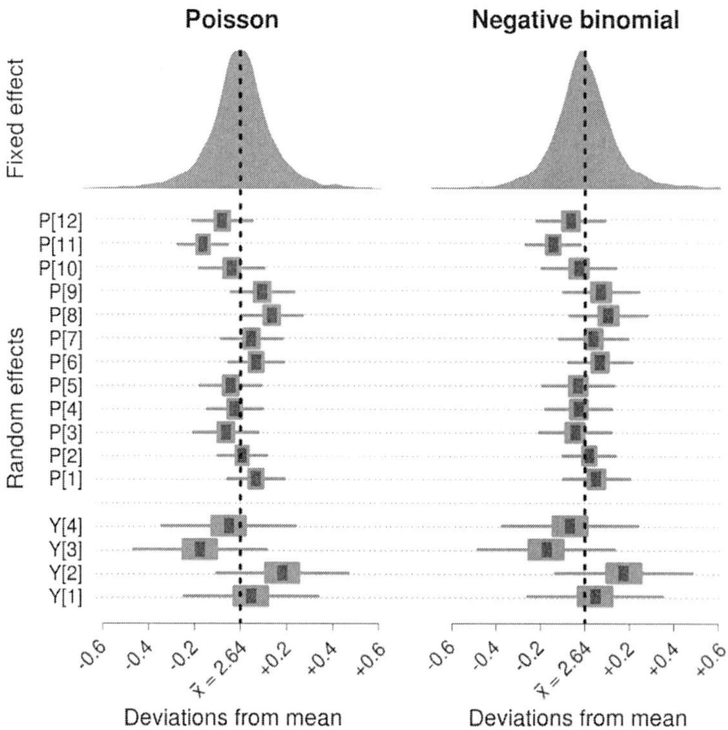

Figure 9.3: Posterior estimates for fixed effects of the Poisson (upper left) and negative binomial (upper right) models of the seed number per fruit. Random deviations due to population identity and year for the Poisson (bottom left) and negative binomial (bottom right)

Both models are very similar in their predictions for these data. The magnitude and distribution of the posteriors of deviations per population (many of them overlapping with zero) provide robust evidence that there is very little variation attributable to the effect of population. The larger magnitude of the deviations by year indicates that there is a stronger temporal variation. The larger average value of the posteriors for the variance among years (*sigmay*) compared with the ones for populations (*sigmap*) confirms this result (Figure 9.4).

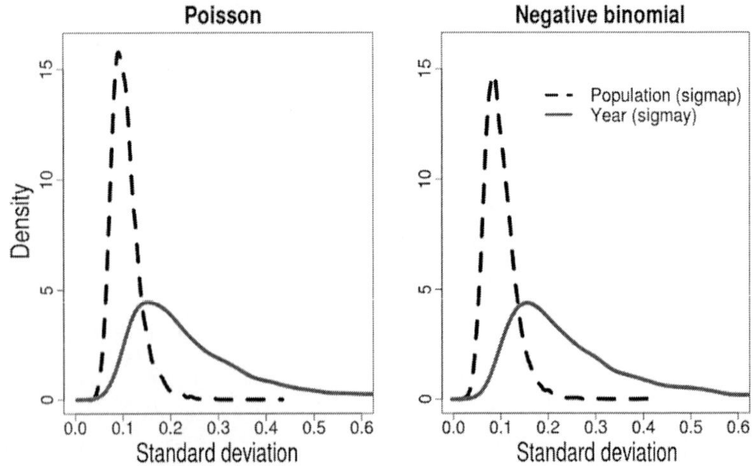

Figure 9.4: Posterior distribution of the variance of the deviations by populations (discontinuous line) and years (continuous line)

We need to decide which model is more likely to explain these data. As recommended, we avoid using the WAIC to compare models with Poisson and negative binomial likelihoods (McElreath 2016). We can instead use three other complementary sources of information. The objective is to find which model is more parsimonious.

The first piece of information is the number of parameters estimated by the model. All else being equal, the fewer the number of parameters needed, the greater are the degrees of freedom and the more parsimonious is the model. Given the same linear predictor, a negative binomial model will always have one less residual degree of freedom than a Poisson model (because of the extra *scale* parameter).

The second source of information that we need to evaluate is the magnitude of the residuals. Smaller values indicate a better model fit. In this case, the mean sum of squares for the negative binomial model is about half of that for the Poisson model (~956 *vs.* ~1863 respectively). If the values were closer, we might feel inclined to opt for the Poisson model. However, this large difference warrants looking at a third source of information: the dispersion statistic. Do not confuse the dispersion *statistic* (see below) with the dispersion *parameter* α for the variance of negative binomial distributions above (Zuur et al. 2015).

$$dispersal\ statsitc\ = \frac{x^2}{residual\ degrees\ of\ freedom} \qquad \text{Equation 9.6}$$

$$x^2 = \sum_{i=1}^{N} \frac{(Y_i - E(Y_i))^2}{var(Y_i)} \qquad \text{Equation 9.7}$$

The dispersion statistic offers a precise way to assess over-dispersion in these models. Over-dispersion occurs when there is greater variability in the data than what is expected from the model. A well-fitted count model should have a dispersal statistic close to 1. In our example, the dispersion statistic for the negative binomial model has a median of 1.007, increasing our confidence in the model (Figure 9.5). In contrast, the dispersion statistic for the Poisson has a median of 2.168, which shows that the model does a poor job of describing the variance of the data.

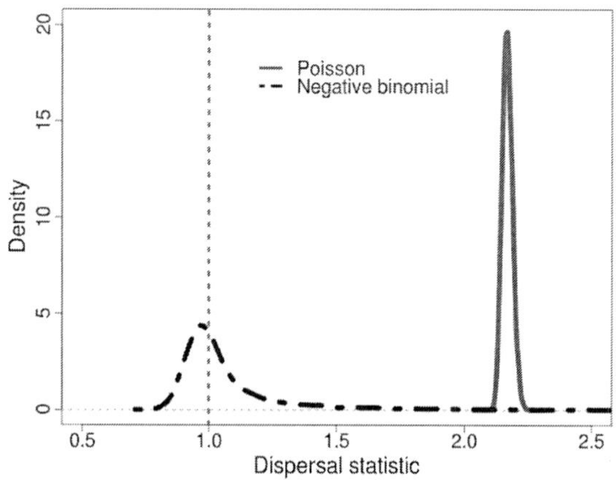

Figure 9.5: Posterior distribution of the dispersal statistic for the Poisson model (discontinuous line) and the negative binomial model (continuous line)

Since the Poisson distribution assumes that the variance is equal to the mean, the model lacks the flexibility to account for any increase in the variance of the data. This helps to explain the much larger sum of squares for the Poisson.

Final considerations

For these data, we favor a negative binomial model because the Poisson model fails to account for the variance of the data. The assumption that the variance is equal to the mean is clearly not supported. Although the mean predictions of both models are very similar, the negative binomial predicts larger variances better suited for these data. In other words, the use of the extra scale parameter seems well justified.

Because of the scale parameter, the negative binomial will often have a dispersion statistic closer to 1 than the Poisson, even if just marginally. Without a technique such as the WAIC, it might seem hard to decide when the use of a Poisson model is preferred. Different authors suggest different cut-offs and approaches to determine what value of the dispersion statistic is acceptable. These are all arbitrary, so we prefer to leave the decision to the reader. In our example, we feel validated in our decision because the distribution of the dispersion statistic for the Poisson model doesn't come close to overlapping 1 (Figure 9.5).

These data indicate that the seed number per fruit in *Hypericum cumulicola* was more consistent among populations than among years (Figure 9.6). There was a large overlap among the posterior distributions of the number of seeds per fruit by population (except population 42) and between them and the posterior distribution of the overall number of seeds per fruit. On the other hand, the posterior distribution of the mean number of seeds per fruit was different among years. One year had its mean higher than the overall mean (1996), one smaller (2014) and two nearer to the overall mean (1995 and 1998). A possible explanation is that temporal variations such as annual precipitation or time-since-fire have a greater influence on the production of seeds than variables affecting spatial conditions, such as abundance of competitors.

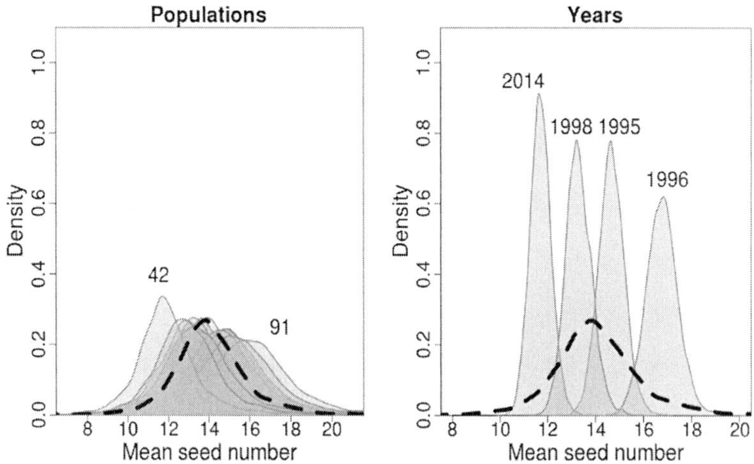

Figure 9.6: Posterior distribution of the mean seed number by population (left) and years (right) for the negative binomial model are indicated as gray polygons. The posterior for the mean overall response is indicated by a dashed black line

CHAPTER X

DATA WITH TOO MANY ZEROS: REPRODUCTIVE BIOLOGY

"If you thought that science was certain – well, that is just an error on your part."
– Richard P. Feynman

Relevant notions

Zeros in data can sometimes exceed what is anticipated by Poisson and negative binomial models (introduced in Chapter IX). This happens when evaluating phenomena are characterized by a mixture of two processes: one defining the conditions for its occurrence and another defining the degree of its manifestation. The ecological drivers for one process can overlap or not with the drivers for the other process. This decoupling, however, means that the non-occurrence of phenomena is a frequent and valid outcome.

This type of data is widespread in biology. As anyone who has performed biological surveys can attest, finding no specimens in a trap, transect, or plot is all too common. The reasons for such absences can be many. For example, when evaluating reproductive outputs, individuals must first be in the reproductive condition before we can estimate the attributes of their progeny. Reproductivity, maturity, and readiness may depend on body condition, ontogeny, and the presence of sexual partners, while the number of descendants may be determined more by biomass and body condition.

It can be argued that separating the analysis of both processes solves the problem, but their separation may not be trivial, as some individuals in the reproductive condition may still not produce progeny. While the occurrence of reproduction is described by binomial distributions, the number of progeny is characterized with Poisson or negative binomial distributions. We want an approach that maximizes the likelihood of the distributions describing both phenomena simultaneously. The goal is to estimate the posterior probability of the mean and uncertainty of the combination of these dependent variables.

Count data with too many zeros may require this type of attention because they may have far more zeros than what would be expected from Poisson or negative binomial distributions. As we already mentioned, there are modeling techniques that can deal with this excess of zeros, but they require an understanding of the nature of those zeros. We adopt the classification described by Martin et al. (2005) to categorize different zeros encountered in ecological studies (Table 10.1).

Structural zeros, those that result from the phenomena considered, are called positive or true zeros, but such a definition is open to discussion (Martin et al. 2005). Zeros that arise from experimental design flaws, such as sampling in impossible species ranges or observer limitations, are called false zeros or false negatives, and we should do our best to avoid them (Zuur et al. 2009). If observer error can be estimated, as in capture-recapture methods, this information can be used to obtain robust count estimates (Kéry and Schaub 2012). Researchers have little control over organismal error, but it can be minimized with better designs and more replicates.

There are several possible ways to analyze zero-inflated count data. Here, we explore two alternative formulations. First, we take a naive and optimistic approach and assume that there was no zero-inflation in our count data. This formulation assumes a Poisson distribution for the likelihood of the model, following the same procedure as in Chapter IX. The second formulation involves a mixed zero-inflated model. In the last model, zeros are assumed as coming from two different processes: the binomial and count processes. The binomial portion estimates the probability of "structural" zeros versus all other types of data (counts and "organismal error" zeros). In this way, it is possible to use different sets of covariates to explain the probability of occurrence and the actual counts. It is important to remark that false zeros are not directly contemplated by these models.

Table 10.1: Classification of the source of zeros modified from Martin et al. (2005)

Types of zeros	Example using counts of individuals	Comments
Structural zeros	Individuals of the study species were not present because the habitat was not suitable for them	These zeros are of interest to the ecological study
Design errors	Individuals were not found because of poor experimental design or incorrect sampling practices	An improvement on the design is necessary
Observer error	Individuals were there but they were not seen	The probability of these zeros needs to be estimated for a correct estimate of the variable of interest
Organism error	The habitat was suitable, but individuals were not there	It is recommended to have enough replications to characterize these errors
Bad zeros	Sampling outside the species range, for example, looking for fish out of water	Requires a clear understanding of the study system

We can use some criteria to select the most informative model. However, as expressed by Zuur et al. (2009), it is always better to include biological knowledge to decide between models. In the example below, we use the mixed zero-inflated type of model because we are convinced of the existence of genuine structural zeros (as suggested in Sujit, Mukhopadhyay and Lu 2006).

-- --

👓 Glasses On 👓

A Poisson model $\mathcal{F}(\rho_i, \alpha; y_i)$, where $y_i = 0,\ 1,\ 2$ ..., is defined by:

$$\mathcal{F}(\rho_{i,}\alpha; y_i) = \left(\frac{\rho_i}{1+\alpha\times\rho_i}\right)^{y_i} \frac{(1+\alpha\times y_i)^{y_i-1}}{y_i!}\ exp\left[\frac{-\rho_i(1+\alpha\times y_i)}{1+\alpha\times\rho_i}\right] \quad \text{(Equation 10.1)}$$

where: $\rho_i = \exp\ (\sum \beta_j \times x_{ij})$, x_{ij} is an element of the covariate matrix X, and β_j are elements of the column vector of unknown parameters. The mean of y_i is given by ρ_i. In this model, α is the dispersion parameter (see Chapter IX). When $\alpha = 0$, the probability model is reduced to a Poisson regression model, whereas $\alpha > 0$ represents data with over-dispersion and is better analyzed by a negative binomial model, and $\alpha < 0$ represents data with under-dispersion.

A mixed zero-inflated generalized Poisson regression model is given by:

$$P(Y = y_i|x_i, z_i) = \mu_i + (1 - \mu_i) \times \mathcal{F}(\rho_i, \alpha; y_i) \qquad y_i = 0$$

$$P(Y = y_i|x_i, z_i) = (1 - \mu_i) \times \mathcal{F}(\rho_i, \alpha; y_i) \qquad y_i > 0$$

(Equation 10.2)

where $\mathcal{F}(\rho_i, \alpha; y_i)$ was defined above and $0 < \mu_i < 1$. The functions $\rho_i = \rho_i(x_i)$ and $\mu_i = \mu_i(z_i)$ satisfy $\log{(\rho_i)} = \sum_{j=1}^{k} x_i \beta_j$, and $logit(\mu_i) = \log(\mu_i \times [1 - \mu_i]^{-1}) = \sum_{j=1}^{k} z_{ij} \Lambda_j$, where z_{ij} is an element of the covariate matrix Z, and Λ_j are elements of the column vector of unknown parameters. The non-negative functions ρ_i and μ_i are, respectively, estimated via log and logit generalized linear models. The counts are assumed occurring in two distinct states. The zeros in the first state occur with the probability $\mu_i + (1 - \mu_i) \times \mathcal{F}(\rho_i, \alpha; y_i)$, when $y_i = 0$ and $\mu_i = P(0)$. Non-zeros in the second state occur with the probability $(1 - \mu_i) \times \mathcal{F}(\rho_i, \alpha; y_i)$, when $y_i > 0$. The mixed distributions lead to the mass function of equation 10.2.

-- -

Practical insights

In this chapter, we don't use observed field data. Instead, we produce computer-generated data from known parameters to help us illustrate the use of zero-inflated models. While it is true that original data have a more complex uncertainty structure, simulated data have several advantages. In this case, because of the way they were created, generated data will only include errors attributable to sampling, thus allowing for a more direct comparison of the two proposed implementations. Second, simulating the data means that we know the data-generating process (which was previously simply assumed). As we know the "actual value" of the expected parameters and their "assigned uncertainty", the evaluation of the results is more transparent.

For the generation of data, we use parameters obtained from a study by Margaret Evans and collaborators (2003; Table 10.2) aimed at understanding the reproductive biology of three rare Florida scrub species. This research involved hand pollinations and observations to study the mating systems and pollination biology of *Eryngium cuneifolium*, *Hypericum cumulicola* and *Liatris ohlingerae* in Archbold Biological Station. To maintain the overall theme of the book, we will only use the information on *H. cumulicola*.

Table 10.2: Manipulations and observed parameters from field data per pollination treatment

	Emas-culated	**Hand pollinated**	**Caged**	**Prop. of fruits mooring**	**Mean # of seeds/mature fruit**
Auto			√	0.068	11.7
Self	√	√	√	0.449	14.1
Cross	√	√	√	0.464	16.4
Control				0.443	14.0

During the reproductive season in 2002, Evans and colleagues marked 80 individuals and emasculated (via the removal of anthers before dehiscence) 40 of them. They hand pollinated with pollen from the same plant (self) the emasculated flowers of 20 plants. The remaining 20 plants with emasculated flowers were pollinated with pollen from a different plant (cross). All the hand-pollinated, emasculated plants, as well as half of those with intact flowers (auto) were individually caged to eliminate access from pollinators. The remaining 20 plants with intact flowers were assigned to a control treatment and left without any cover. They counted the number of seeds developed by flowers in the experiment.

Evans at al. (2003) split their results into fruit set and seed set. In our numerical experiment, we merge these data and compare seed counts among treatments. We use their estimates to build a predicted realization of the number of seeds per each treated flower of 20 flowers from 20 imaginary plants per treatment and analyze their differences. We assume random sampling with binomial and Poisson likelihoods during data generation. The first one defines fruit ripening and the second seed number per mature flower. However, the count is by flower treated. The number of extra zeros in these data is notable (Figure 10.1). The process to generate simulated data with an *R* code is in Box 10.1. Note that the actual counts of seeds per pollinated plant will change in different realizations of this simulation.

Box 10.1: Code for data generation

```
## read the original field data from file
fruit_data <- read.csv("HCfruits.csv", header=T)
seed_data <- read.csv("HCseeds.csv", header=T)

## prepare arrays for saving the data
T_P <- array(0, c(2,4))  # averages
zero <- array(0,c(20,4)) # extra zeros per plant

colnames(T_P) <- c("auto", "self", "cross", "control")
rownames(T_P) <- c("fruit_set","seeds")

## generate the extra zeros
  s <- 0
  for (j in 1:4){

T_P[1,j]<-
round(mean(fruit_data$nofull[fruit_data$treat==j+s]/fru
it_data$nostruct[fruit_data$treat==j+s]),3)

T_P[2,j]<-
round(mean(seed_data$NOFULL[seed_data$treat==j+s]/seed_
data$X_FREQ_[seed_data$treat==j+s]),0)

zero[,j] <- 20-20*(rbinom(20,20,T_P[1,j])/20)
s <-1
   }

## generate seed counts and group them with the extra
zeros per treatment
  yes_o <- yes_c <- yes_s <- yes_a <- NULL
```

```
  for(j in 1:20){
  yes_a <- c(yes_a,rpois(20- zero[j,1],T_P[2,1]),
  rep(0,zero[j,1]))
  yes_s <- c(yes_s,rpois(20-zero[j,2], T_P[2,2]),
  rep(0,zero[j,2]))
  yes_c <- c(yes_c,rpois(20-zero[j,3],T_P[2,3]),
  rep(0,zero[j,3]))
  yes_o <- c(yes_o,rpois(20-zero[j,4],T_P[2,4]),
  rep(0,zero[j,4]))
  }

## finish compiling the simulated data
id <- rep(1:80, each = 20, len = 1600)
treat <- rep(1:4,each=20*20,len=20*20*4)
cg_seed
<- cbind(treat,rep(id,4),c(yes_o,yes_c,yes_s,yes_a))
colnames(cg_seed) <- c("treat","plant","seed")
cg_seed <- as.data.frame(cg_seed)
```

Model building

We evaluate two models to analyze the generated data and test model efficacy, one model with zero inflation and another without. Because of the data generation process, we expect that not all the flowers will develop as mature fruits, and these partially developed fruits contribute extra zeros to seed counts per studied flower.

The *Stan* code for the Poisson model with no zero inflation, Model 10.1, uses model specifications like those indicated in Chapter IX, but including fixed effects of pollination treatment and random effects by plant (instead of population and year). Our proposed mixed zero-inflated generalized binomial/Poisson model, Model 10.2, also evaluates the amount of variation estimated among plants, assuming random effects, and the fixed effects of pollination treatment. Its *Stan* code is detailed in Box 10.2. Remember that you need an *R* code to connect with this *Stan* code (see the annotated code on the book's website).

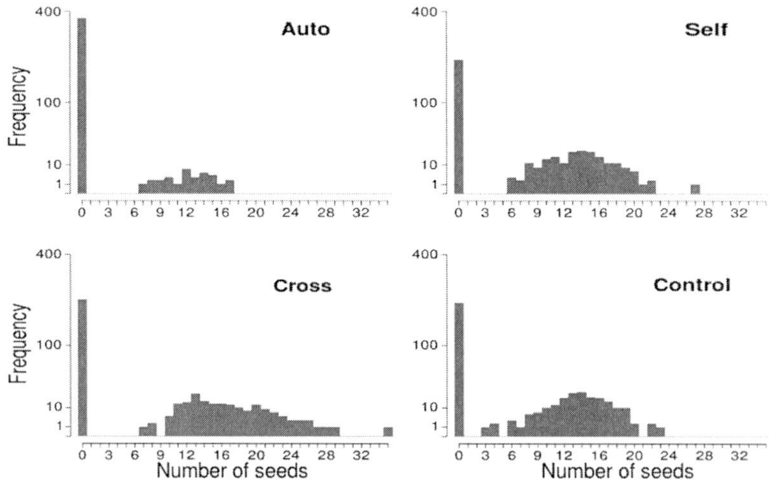

Figure 10.1: Histogram of a realization of the simulated pollination data. The y-axis has been square-root-scaled to help visualize the data

Box 10.2: *Stan* code for Model 10.2, a mixed zero-inflated generalized binomial/Poisson linear regression model

---

```
// Data declaration
data{
    int<lower=1> N;
    int<lower=1> N_plant;
    int seed[N];
    int plant[N];
    real self[N];
    real cross[N];
    real control[N];
}
// Parameter declaration
parameters{
    real a;
    real b1;
    real b2;
    real b3;
    real z1;
    real z2;
    real z3;
```

```
    real q;
    vector[N_plant] p1;
    real<lower=0> sigmap1;
    vector[N_plant] p2;
    real<lower=0> sigmap2;
}
```

// Model declaration including prior specification

```
model{
    vector[N] lambda;
    vector[N] pbar;
    sigmap2 ~ cauchy( 0 , 10 );
    p2 ~ normal( 0 , sigmap2 );
    sigmap1 ~ cauchy( 0 , 10 );
    p1 ~ normal( 0 , sigmap1 );
    b3 ~ normal( 0 , 10 );
    b2 ~ normal( 0 , 10 );
    b1 ~ normal( 0 , 10 );
    q ~ normal( 0 , 100 );
    z3 ~ normal( 0 , 10 );
    z2 ~ normal( 0 , 10 );
    z1 ~ normal( 0 , 10 );
    a ~ normal( 0 , 100 );
    for ( i in 1:N ) {
        lambda[i] = q + z1 * self[i] + z2 * cross[i] +
z3 * control[i] + p2[plant[i]];
        lambda[i] = exp(lambda[i]);
    }
    for ( i in 1:N ) {
        pbar[i] = a + b1 * self[i] + b2 * cross[i] + b3
* control[i] + p1[plant[i]];
        pbar[i] = inv_logit(pbar[i]);
    }
    for ( i in 1:N )
    if (seed[i] == 0)
      target += (log_sum_exp(bernoulli_lpmf(1|pbar[i]),
        bernoulli_lpmf(0|pbar[i]) +
poisson_lpmf(seed[i]|lambda[i])));
    else
      target += (bernoulli_lpmf(0|pbar[i]) +
poisson_lpmf(seed[i]|lambda[i]));
```

```
}
```

// Declaration of generated quantities

```
generated quantities{
    vector[N] lambda;
    vector[N] pbar;
    real dev;
    dev = 0;
    for ( i in 1:N ) {
        lambda[i] = q + z1 * self[i] + z2 * cross[i] +
z3 * control[i] + p2[plant[i]];
        lambda[i] = exp(lambda[i]);
    }
    for ( i in 1:N ) {
        pbar[i] = a + b1 * self[i] + b2 * cross[i] + b3
* control[i] + p1[plant[i]];
        pbar[i] = inv_logit(pbar[i]);
    }
    for ( i in 1:N )
    if (seed[i] == 0)
      dev = dev + (-
2)*(log_sum_exp(bernoulli_lpmf(1|pbar[i]),
        bernoulli_lpmf(0|pbar[i]) +
poisson_lpmf(seed[i]|lambda[i])));
    else
     dev = dev + (-2)*(bernoulli_lpmf(0|pbar[i]) +
poisson_lpmf(seed[i]|lambda[i]));
}
```

Model comparison using the WAIC is not recommended for models having different likelihood distributions, but the summary of Model 10.2 (Table 10.3) shows how well the model predicts both the probability of fruit mooring and seed counts for all treatments. The posterior distributions of the average seed counts per treatment of both models (Figure 10.2) demonstrates the stark contrast between the two models. The posterior distribution for our zero-inflated model overlapped the observed seed counts (for mature fruits) in all treatments. Conversely, the Poisson model without zero inflation grossly underestimated seed counts.

Table 10.3: Observed and estimated predicted probability of mooring and seed counts. Predicted values and coefficients after the zero-inflated Poisson model (Model 10.2). The predicted values will be different among realizations of the generated data and the model implementations

Coefficient	Mean	Std Dev	Mooring	
Binomial			**Observed**	**Predicted**
Intercept (Auto)	2.86	0.15	0.068	0.055
Self	-2.59	0.20	0.449	0.434
Cross	-2.62	0.20	0.464	0.442
Control	-2.75	0.20	0.443	0.473
Sigma among plants	0.49	0.06		
Poisson			**Seed count**	
Intercept (Auto)	2.46	0.04	11.70	11.68
Self	0.18	0.04	14.10	14.00
Cross	0.31	0.04	16.40	15.85
Control	0.20	0.04	14.00	14.28
Sigma among plants	0.09	0.01		

The miscalculation of the first model was most extreme for the *auto* treatment, where the lack of pollinators (humans or otherwise) led to a greater excess of zeros. The binomial portion of the zero-inflated model correctly accounted for this, estimating a lower probability of mooring for fruits in the *auto* treatment (Figure 10.3).

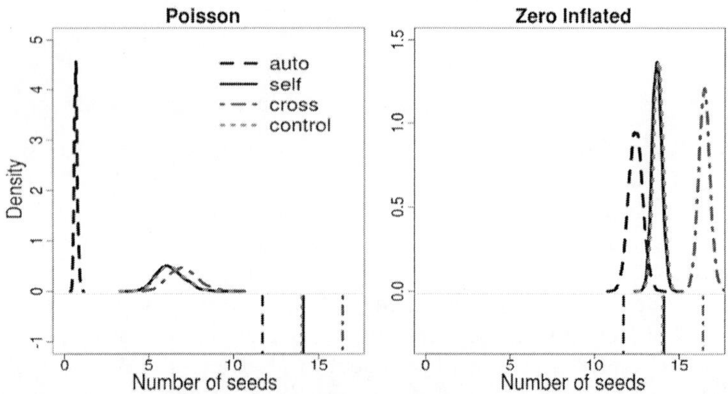

Figure 10.2: Predicted posterior average seed counts per pollinated flower for each treatment after a linear Poisson model (left) and a mixed zero-inflated Poisson model (right). Vertical lines underneath show the observed average number of seeds after accounting for the maturation of fruits

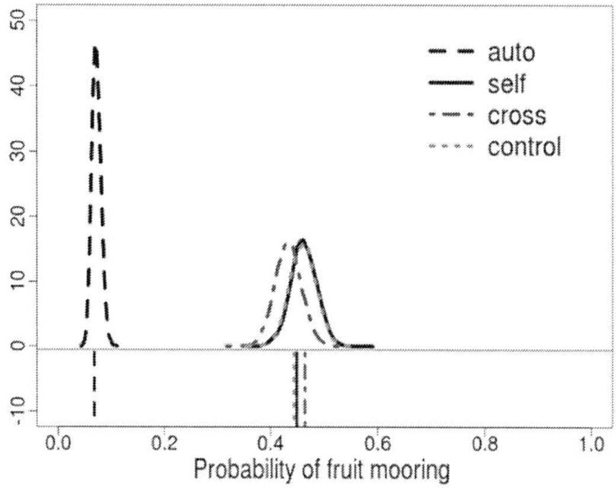

Figure 10.3: Predicted posterior probability of fruit mooring in each treatment obtained from the mixed zero-inflated Poisson model. Vertical lines underneath show the observed proportion of mooring fruits

Final considerations

Given the heavy computational investment that these models require, we might wonder, why not simply model counts in mature fruits? One reason, as already mentioned, is that zeros are sometimes valid counts. Distinguishing between different types of zeros in our data might not be straightforward. When we suspect the occurrence of structural zeros in count data, zero-inflated models are robust tools to assess statistical patterns of interest. They allow us to incorporate other knowledge to help estimate structural zeros.

In our example, we were able to recover reasonable estimates of the seed counts from generated data using information observed by Evans and colleagues (2003). Zero-inflated models have been successfully implemented to deal with count data including too many zeros in ecology and many other disciplines (references in Famoye and Singh 2006). The study of Evans et al. (2003) indicated little evidence of asexual reproduction and a strong dependence on pollinators for reproduction in *Hypericum cumulicola*, which is a hermaphroditic plant with slightly asynchronous maturation and a strong spatial separation of male and female organs in each flower (herkogamy).

CHAPTER XI

MODEL SELECTION FOR MIXED MODELS: EFFECTS OF SIZE, REPRODUCTION, AND TIME-SINCE-FIRE ON SURVIVAL

"Statisticians are paid to be skeptics; they are the conscience of science."
– *Judea Pearl*

Relevant notions

To discuss model selection, and particularly its role in advancing our knowledge, we must retrace our steps to a notion introduced in the first chapter: models are partial descriptions of reality. In view of this, we might wonder how reasonable it is to choose between such models. We can begin by realizing that even the language we use puts a constraint on our ability to perceive reality. Yet, as Nietzsche famously stated, "we cease to think if we do not want to do it under linguistic constraints." Likewise, we would cease all scientific progress if we limited ourselves to the search for absolute certainty.

Science advances through the evaluation of hypotheses – proposed explanations based on previous information used to initiate further research. We concord with Betts et al. (2021) that such evaluation is key to enable the search for mechanisms, reduce bias, and increase reproducibility. We develop models because they allow us to express ideas about data variability, even if those ideas will never be true. This is one of the main reasons that we discourage the classical null hypothesis testing. We believe we can do better than simply rejecting an idea that is wrong by definition.

When assessing the causes of phenomena variation, it is prudent to propose competing explanations. The ideas in these hypotheses are better expressed as mathematical equations in models formalizing the expected relationships among the variables. Model selection is an approach that evaluates the relative support of these models (Burnham and Anderson 2002). It has gained wide acceptance in ecological and evolutionary research, and for good reason. As it ranks and weights the relative information among proposed models, it allows us to learn something from each model. To minimize the risk of spurious discoveries, however, model selection must not be constructed as data dredging (comparing every possible model). Rather, it is best to critically select models reflecting genuine expectations based on knowledge (Burnham and Anderson 2002).

Model selection uses maximum likelihood scores as a measure of agreement between the models and the data, using the number of parameters in each model as a measure of model complexity. The Akaike information criterion penalizes the number of parameters in the model to favor those balancing maximum data agreements with a minimum number of parameters. The WAIC is an extension of the AIC applicable to Bayesian models (Watanabe 2013, McElreath 2016; see Chapter VI).

Practical insights

Survival is a critical, vital rate. Alongside reproduction, it determines the fitness of individuals – the engine of evolution. It is key to understanding population trends and the composition of communities, as well as a host of ecological relations. Variation in survival can be difficult to study, as it depends on a multitude of factors. For example, there is extensive support for ontogenetic variation in survival, with many species having the highest endurance at intermediate ages and sizes (Jones et al. 2013). A general trade-off between reproduction and survival is also well known. In Chapter VI we provided evidence of how the number of reproductive structures of *Hypericum cumulicola* is negatively related to plant survival. In this chapter, we evaluate the relevance of three other variables – plant height, the number of stems, and time-since-fire – to explain survival variation in the same species (Quintana-Ascencio et al. 2003, 2018 and 2019).

Every year, with the start of the rainy season in Central Florida (around mid-May), most of the recently developed vegetative stems of *H. cumulicola* perish, and the few remaining ones develop as indeterminate flower stalks. These will bear all the flowers and fruits for the year. The number of stems that develop into reproductive stalks depends primarily on plant size. In winter, these stalks die, and new ones develop the following year. So, when evaluated during the reproductive season (June-September), the number of stems and plant height represent different ways to describe plant size. Simultaneously, they are also good indicators of reproductive output (see Chapter VII).

As we suspect a certain redundancy (in statistics evaluated as collinearity) between these two variables in their effect on survival, we must be vigilant that they do not artificially inflate the variance of the estimated regression. We can begin with exploring the correlation between them by obtaining a Pearson correlation coefficient. For the data at hand, we obtain $\rho \cong 0.09$, which does not indicate any large redundancy between these variables. In this instance, we feel comfortable enough with the result to continue using both in our assessment. However, a lack of correlation is not always an indication of a lack of collinearity. In some cases, multicollinearity might exist between three or more predictor variables even if no single pair is highly correlated. We will revisit these concepts at the end of this chapter.

Plant size affects the resource budget for any plant with potential consequences on its persistence. The models we propose will allow us to evaluate the relative effect of height and the number of stems and to describe the form of their effect on survival.

Standing *Hypericum cumulicola* is frequently killed or damaged by fire but the species benefits from recurrent fire that reduces competition with dominant shrubs, releases nutrients and increases water availability. We expect changes in survival of this small herb with time-since-fire as competition with dominant shrubs increases and available nutrients and water decrease.

🕶 Glasses On 🕶

The effect size of any categorical variable in linear models is often evaluated with binary combinations of dummy variables which we should specify in the data matrix. This type of specification has the advantage that it is orthogonal and avoids redundancy in the information of the coefficients allowing them to be estimated independently. First, we identify a reference level for this variable. The posterior prediction of the dependent variable associated with this reference categorical predictor level is estimated with the intercept of the model. All the other levels of this categorical variable have their own column. In the respective column, a value of one is assigned to the cases associated with this level and a value of zero to all the other cases. There will be *m-1* columns, *m* being the number of levels for this variable. Then this dummy variable multiplies times one the coefficient of this effect size when the cases are associated with the focal level of the variable and times zero all other cases. Notice that using dummy variables does not allow for a direct estimation of all the comparisons among levels for any independent categorical variable with more than two levels. The calculation of the posterior values of the dependent variable associated with the non-reference levels and other effect sizes not estimated directly can be calculated as derived variables. This requires the implementation of algebraic sums of the intercept and the coefficients of the non-reference levels. For example:

$$Y_{ji} = intercept + \beta_j \times dummy_j \qquad \text{(Equation 11.1)}$$

where Y_{ji} is the posterior associated with the level *j* of a categorical variable and β_j is the coefficient estimating the difference in effect between this focal level and the reference levels of this categorical variable.

Model building

We use a model selection approach to assess the relative importance of these variables and their two-way interactions. To simplify, in these models we do not include any three-way interaction or random effects of site and year. In a real application these elements deserve more attention. We propose six hypotheses evaluated with ten models (Table 11.1): (1) Survival increases with height as individuals mature and accumulate resources; (2) Survival increases with height as individuals mature and then decrease due to senescence; (3) Survival decreases with the number of reproductive stems consistent with a trade-off between these vital rates (see Chapter VII); (4) Survival decreases with time-since-fire as access to resources declines due to an increase of competitors; (5) There are additive effects on survival among these variables acting independently; and (6) There are interactive effects on survival among these three variables due to possible compromises on how resources affect survival.

Here, we specify time-since-fire (TSF) and number of stems as continuous variables. In the book website we implement the models with TSF as categorical variable. In those models, we chose "recently burned" as a reference and using dummy variables (see above) we measured the difference in survival between this condition and two other time-since-fire levels (intermediate and long-unburned). In our models, we obtain coefficients indicating the difference in odds of survival between recently burned and intermediate time-since-fire, and between recently burned and long time-since fire, but there is no coefficient in these models for the difference in effect between intermediate and long-unburned time-since-fire. The last difference must be calculated algebraically as a derived quantity from the estimates in the model.

As in Chapter VII, the likelihood distribution in these models is given by:

$$Survival_i \sim Bin(\pi_i)$$
$$E(survival_i) = \pi_i$$
$$\text{variance}\ (survival_i) = \pi i * (1 - \pi i)$$

The link function for the logistic model is the logit of π. The inverse of the link function is the estimate π_i, which, in this case, represents the probability of survival as a function of *X* (the covariates):

$$\pi_i = \frac{e^{\beta_0+\beta_1 X}}{1+e^{\beta_0+\beta_1 X}}$$

$$\text{Logit}\ (\pi) = \eta \qquad \text{(Equation 11.2)}$$

where η is a predictor function of the covariates (*X*), and β_0 and β_1 are their model coefficients:

$$\ln\left(\frac{\pi_i}{1-\pi_i}\right) = \eta = \beta_0 + \beta_i X_i$$

(Equation 11.3)

An example of its implementation using the most likely model of the chosen set (see below) is in Box 11.1.

Box 11.1: Simplified implementation in *R* and *Stan* for Model 11.3

```
// likelihood
surv_fin ~ binomial_logit(1,p),

// model specification
For (i in 1:N){
 p[i] = a  + b* ln(height)+ bb* ln(height)2 +  c*TSF +
 d*stems + bc* ln(height)*TSF + bd* ln(height)*stems +
 cd*TSF*stems   +   bbc*   ln(height)2*TSF   +   bbd*
 ln(height)2*stems,

// prior of intercept
    a ~ dnorm(0,100),
// prior of coefficient for ln(height)
    b ~ dnorm(0,1),
// prior of coefficient for ln(height)2
    bb ~ dnorm(0,1),
// prior of coefficient for TSF
    c ~ dnorm(0,1),
// prior of coefficient for rep stems
    d ~ dnorm(0,1)
// prior of interaction of ln(height) with TSF
    bc ~ dnorm(0,1),
// prior of coefficient of ln(height) with stems
    bd ~ dnorm(0,1),
// prior of coefficient of TSF with stems
    cd ~ dnorm(0,1),
// prior of coefficient ln(height)2 with TSF
    bbc ~ dnorm(0,1)
// prior of coefficient ln(height)2 with stems
    bbd ~ dnorm(0,1)
```

Table 11.1: Model equations proposed to evaluate the set of hypotheses. Symbols: h = height; s = stems; TSF = time-since-fire; β0 = intercept of the model; η is the logit of the probability of survival; * indicates a model interaction and implicitly all lower terms of variables involved. Negative values under hypotheses indicate that the equation does not include a critical term for this hypothesis and can be used as a reference for the relevance of the factor evaluated with a corresponding model that includes it

Model	Model equation	Hypotheses
11.1	$\eta = \beta_0 + h * s + h * TSF + s * TSF$	1, 3, 4, 5, 6
11.2	$\eta = \beta_0 + h + s + TSF$	1, 3, 4, 5, -6
11.3	$\eta = \beta_0 + \mathrm{h} * s + h * TSF + h^2 * s + h^2 * TSF + TSF*s$	2, 3, 4, 5, 6
11.4	$\eta = \beta_0 + h + h^2 + s + TSF$	2, 3, 4, 5, -6
11.5	$\eta = \beta_0 + h + h^2 + TSF$	2, -3, 4, 5, -6
11.6	$\eta = \beta_0 + h + h^2 + \mathrm{s}$	2, 3, -4, 5, -6
11.7	$\eta = \beta_0 + h * TSF + h^2 * TSF$	2, -3, 4, 6
11.8	$\eta = \beta_0 + \mathrm{h} * s + h^2 * s$	2, 3, -4, 6
11.9	$\eta = \beta_0 + h + h^2$	2, -3, -4, -5, -6
11.10	$\eta = \beta_0 + h$	1, -3, -4, -5, -6

To better understand how the results from multiple models can help to inform our conclusions, it is useful to inspect their coefficients (Table 11.2 – parts 1 and 2). The intercept in these equations is merely a reference point, and it does not have any real-world interpretation (after all, there is no such thing as a plant with no height and no stems). Model 11.3, for instance, predicts approximately 0.68-0.73 (for the mean expectation: 1/(1+exp(-0.89))), for the probability of survival for the reference point in recently burned habitats. As you can notice, the intercept is highly variable among models – this is because the conditions that each intercept represents are different. For example, in the most likely Model 11.3 (see Table 11.2 – part 2) the intercept is the mean estimate of the odds of survival when the height and number of reproductive stems are assumed to be zero, and the plants belong to the chosen reference level of time-since-fire (in these models, "recently burned). In Model 11.10, conversely, the intercept only contemplates the value for height of zero, with the effects of the other variables considered as part of the uncertainty.

Table 11.2 (part 1): Estimated mean and standard deviation (below each mean coefficient and in *italics*) for the coefficients of the independent variables. Coefficients: a: intercept; b: height; c: time-since-fire; d: # of stems, and their respective two-way interactions. The order of models is descending with their WAIC weight

model	a	b	c	d	bc	bd	cd
11.3	0.89	0.81	-0.01	-0.04	-0.01	-0.12	0.001
	0.12	*0.11*	*0.005*	*0.071*	*0.004*	*0.05*	*0.0001*
11.4	0.91	0.39	-0.01	-0.12			
	0.06	*0.05*	*0.001*	*0.009*			
11.8	0.72	0.62		-0.10		-0.11	
	0.10	*0.09*		*0.05*		*0.05*	
11.6	0.80	0.39		-0.12			
	0.06	*0.05*		*0.009*			
11.1	1.34	-0.07	-0.03	-0.07	0.004	-0.03	0.002
	0.09	*0.03*	*0.004*	*0.04*	*0.001*	*0.01*	*0.001*
11.2	1.21	-0.05	-0.01	-0.13			
	0.05	*0.02*	*0.001*	*0.01*			
11.7	0.84	0.62	-0.01		-0.02		
	0.09	*0.09*	*0.005*		*0.005*		
11.5	0.78	0.37	-0.01				
	0.06	*0.05*	*0.001*				
11.9	0.65	0.37					
	0.06	*0.05*					
11.10	0.99	-0.15					
	0.05	*0.01*					

The coefficients associated with the linear response of survival to the logarithm of height (b) can be grouped into three sets. The first group of coefficients associated with height are smaller and negative (-0.15 to -0.5) and occur in models where there is no quadratic coefficient for the variable (Models 11.1, 11.2, and 11.10). These coefficients recognize the general decline in survival with height when forced to ignore the rising trend in survival for plants smaller than 5 cm (Figure 11.1). This underscores the importance of inspecting the observed data visually when proposing a modeling set. The second set of coefficients of height includes positive values around 0.37 to 0.39. They are associated with models including the quadratic response to height but no interacting effects (Models 11.4, 11.5, 11.6, and 11.9). Coupled with the negative coefficient of the quadratic portion (bb), it better describes the initial increase in survival for small plants and the subsequent decline for plants larger than 5 cm. The final group shows even higher positive values for these coefficients (0.62-0.81). They represent the linear response to height for models with a quadratic component and interactions between some variables (Models 11.3, 11.7, and 11.8). The stronger effect of height in these models is a consequence of the overall higher survival of plants with fewer reproductive stems in recently burned habitats compared to any other time-since-fire.

Table 11.2 (part 2): Estimated mean and standard deviation (below each mean coefficient and in *italics*) for coefficients involving the quadratic component of height. Coefficients bb for height^2 and its interactions with time-since-fire (bbc) and the number of stems (bbd). Also, summary values after model selection via the WAIC

model	bb	bbc	bbd	WAIC	dWAIC	Weight
11.3	-0.22	0.004	0.03	22921	0.0	1.0
	0.02	*0.001*	*0.009*			
11.4	-0.10			22971	49.9	0.0
	0.01					
11.8	-0.16		0.02	22991	69.9	0.0
	0.02		*0.009*			
11.6	-0.10			23002	80.8	0.0
	0.01					
11.1				23015	93.9	0.0
11.2				23047	126.4	0.0
11.7	-0.20	0.005		23107	186.7	0.0
	0.02	*0.001*				
11.5	-0.12			23159	238.2	0.0
	0.01					
11.9	-0.12			23202	281.3	0.0
	0.01					
11.10				23305	384.3	0.0

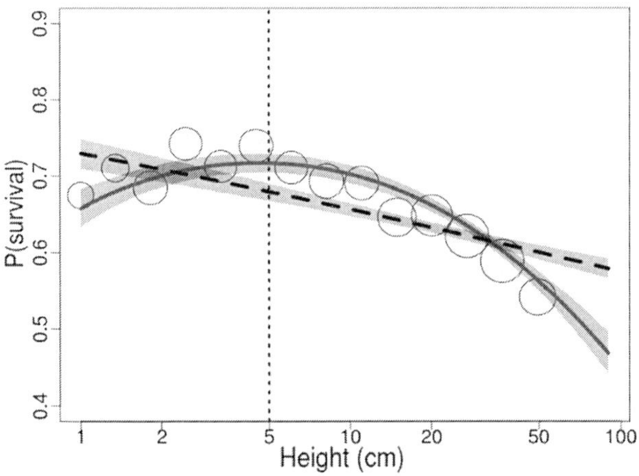

Figure 11.1: Plot of survival as a function of height and mean predictions (95% credibility intervals) for Model 11.9 (solid gray line) and Model 11.10 (dashed black line). The open circles are proportions of observed alive plants of data binned using regular height intervals. The size of each circle is proportional to the sample size in each bin (logarithmically scaled since it ranges from 114 to 2379). The vertical line represents approximate height when survival is highest

As already hinted, in a quadratic polynomial, the combination of a positive linear coefficient (b) with a negative quadratic term (bb) defines an equation for a hump-shaped model (Figure 11.1). For the models with interactions the quadratic term was also more negative, indicating the sharper rise and fall in survival probability with increasing height in a recently burned habitat. The negative coefficients for time-since-fire are relatively consistent among models (-0.01 to -0.03) while the negative coefficients for the number of reproductive stems are more variable. Larger negative values for the coefficients of stems (-0.12 to -0.13) are associated with models without interactions with height (Models 11.2, 11.4, and 11.6), while smaller negative values (-0.04 to -0.10) are in models with interactions (Models 11.1, 11.3, and 11.8). These later coefficients illustrate changes in probability of survival as a function of the number of stems for plants of different heights. The coefficients for the interaction between reproductive stems and height are more negative for the model involving a quadratic height term. Other coefficients indicating interactions were relatively more consistent among models.

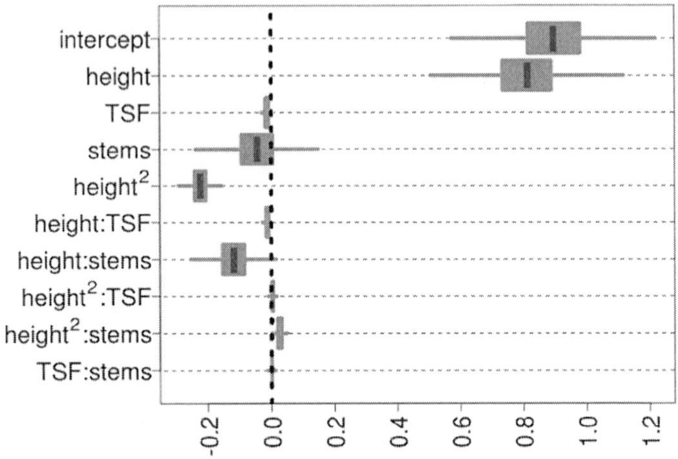

Figure 11.2: Plot of coefficients of Model 11.3 (mean $\pm$ 95% credibility interval) explaining the variation of survival as a function of plant height (log transformed), the number of reproductive stems and time-since-fire

Among the proposed set, Model 11.3 conveys most information. It has the lowest WAIC and a very high weight as a likely explanation (Table 11.2 – part 2). Model 11.3 includes all the variables considered and their two-way interactions. It also provides support for a unimodal change (hump-shaped) in survival with increasing size (hypothesis 2), and a negative association of survival with the number of flowering stems (hypothesis 3) and time-since-fire (hypothesis 4). It also supports an interactive effect of these variables (hypothesis 6).

Our top model (11.3) also reveals the higher uncertainty in the prediction of mean survival for small plants with more than 3 stems – reflecting the rarity of such plants in the sample data (Figure 11.3). Except for the coefficient of reproductive stems whose 95% credibility interval overlaps zero (-0.045 to 0.11), all other coefficients are clearly different from zero.

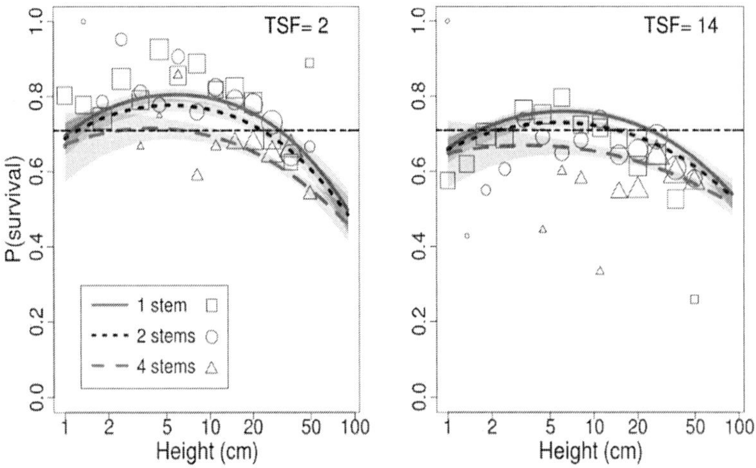

Figure 11.3: Plot of survival as a function of plant height for different times-since-fire and numbers of reproductive stems. TSF = 2 (left plot) and 14 (right plot). Reproductive stems = 1 (solid gray line and squares), 2 (dotted black line and circles), and 4 (dashed gray line and triangles). Shades represent 95% credibility intervals. Sizes of dots are logarithmically proportional to the sample size of plants binned by height and TSF (+/- 5 years)

The next possible model (11.4) also included all the variables assessed but no interactions, and this is approximately 50 times less likely than Model 11.3 as the process generating our data. The model in third position includes interactions between height (also with a quadratic term) and the number of reproductive stems, confirming the relevance of the trade-off between reproduction and survival. These two last models provide substantially more information than the model only including information on plant height (Table 11.2 – part 2).

Before moving on, we should revisit our assumption that any potential collinearity between our predictor variables did not cause misleading model inferences. The low correlation between height and number of stems does not guarantee a lack of collinearity in our model. To better assess the validity of our assumption, we can use a quotient known as the variance inflation factor (VIF), which quantifies the severity of multicollinearity (see below). It estimates the extra variance in the estimates for the effect of each independent variable caused by its redundance with other variables. For our three predictor variables, the mean VIF values were approximately: 1.21 for height, 1.01 for time-since-fire, and 1.22 for number of stems (Figure 11.4). As you guessed, a value close to one is an indicator of minimum redundancy. There are different rules-of-thumb for what constitutes an acceptable VIF value. In reality, these cut-offs are artificial, and careful consideration of the data and the scale of analysis is needed. The low values obtained in our case, however, support our initial assumption.

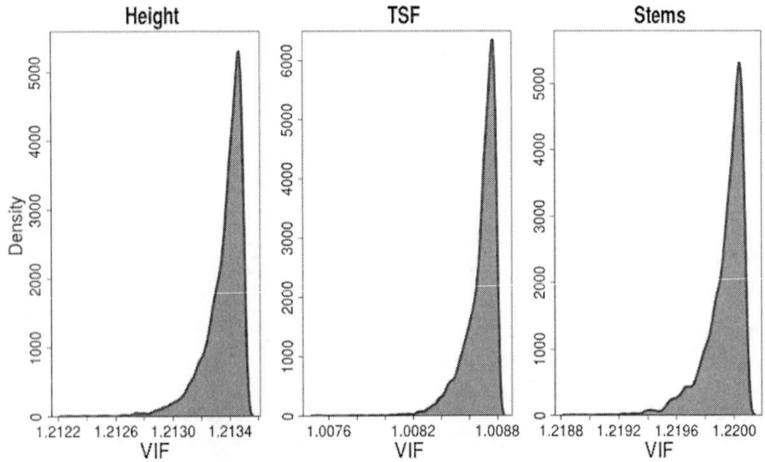

Figure 11.4: VIF estimates for each of our predictor variables: Logarithm of height (left), time-since-fire (center) and number of reproductive stems (right). Estimates are presented as distributions because they were estimated using MCMC estimation

-- --

🕶 Glasses On 🕶

Correlation and collinearity can be related to each other but mean different things. Simply put, correlation refers to the association between any two given variables. There are different ways to measure it, depending on the nature of the data. For instance, the popular Pearson correlation coefficient (ρ_{xy}) measures the linear association between two variables, expressed as a value between -1 (negatively correlated) and 1 (positively correlated). It is calculated as:

$$\rho_{xy} = \frac{cov(x,y)}{\sigma_x \times \sigma_y} \quad \text{(Equation 11.4)}$$

It is worth noting that variables are not always associated linearly. For instance, if two variables are related to each other as a parabola, the Pearson coefficient for these variables in the natural scale will be close to zero – effectively unable to identify the relationship.

The estimate of correlation is useful. For example, when running a linear regression, we expect a linear association between any predictor X_i and the evaluated response Y. A strong correlation between an independent and a dependent variable is exactly what we want. When a strong correlation occurs between two predictor variables in a model, however, we might run into a problem called collinearity. When two predictor variables change together, it becomes hard to separate the magnitude of their specific effect on the dependent variable. In other words, the two variables are somewhat redundant as predictors.

Collinearity affects the estimated value of the coefficients, inflating their variances, and potentially leading to improper model inferences. While a strong correlation between two predictors indicates a potential collinearity problem, collinearity can also occur between multiple variables (3 or more) with no actual pair being highly correlated. These cases are sometimes referred to as multicollinearity.

A good way to quantify potential collinearity, even between multiple predictors, is by estimating the variance inflation factor (VIF) of each predictor X_i. The formula to calculate the VIF is:

$$VIF_i = \frac{1}{1-R_i^2} \qquad \text{(Equation 11.5)}$$

where R_i^2 is the coefficient of determination of X_i, calculated by running a linear regression where X_i is a function of the remaining predictors. For instance, for X_i (for i different from the index of other predictor variables), the regression would be:

$$X_i = \beta_0 + \beta_1 \times X_1 + \beta_1 \times X_2 + \ldots \beta_n \times X_n + \varepsilon \qquad \text{(Equation 11.6)}$$

R^2 is the proportion of the variance in this dependent variable (X_1) predicted by the other independent variables. When R^2 is close to 1, the VIF will be very high, and when it is close to 0, the VIF will be close to 1.

As we mentioned earlier, different authors propose different cut-offs for what represents an acceptable VIF quotient. In general, the closer the value to one, the better. However, the actual value used is arbitrary, and it will depend on such things as the nature of the data, sample size and scale of the analysis – generally speaking, collinearity is less of a problem with larger sample sizes and in carefully designed experimental studies.

-- --

Final considerations

Our goal in this chapter was to exemplify how model selection can be used to evaluate alternative hypotheses. We advise against the passive approach of using model selection to simply identify the "best" combination of variables from all possible permutations. Instead, we want to encourage taking an active role in exploring models that allow the testing of relative support for carefully chosen hypotheses. Simultaneous inspection of the models in the set can help further our understanding of the phenomena in question.

This approach allowed us to learn about possible senescence and reproductive trade-offs in *Hypericum cumulicola* and changes in its persistence with time-since-fire. This information can be re-evaluated in the context of current available information to guide management and to advise life-history theory.

Finally, we want to stress the tension between inference and available data. We consider models as incremental steps, allowing us to identify promising relationships that can be evaluated with future research. Statistical models are constrained by the data used to evaluate them. Careful consideration of the information obtained can indicate areas requiring re-examination with better sampling designs, more data, and, in many cases, experiments that can overcome the current weaknesses of available data.

CHAPTER XII

SPATIAL AND TEMPORAL VARIATION: EFFECTS OF FIRE AND HABITAT STRUCTURE ON REPRODUCTION

"I can be changed by what happens to me. But I refuse to be reduced by it."
– Maya Angelou

Relevant notions

A central assumption that often remains unspoken in scientific pursuits is that our work is performed at the proper scale. As Simon Levin (1992) famously stated, however, ecological phenomena are the consequence of factors interacting at multiple scales (temporally, spatially, and organizationally). Therefore, no single scale is appropriate to all cases. Ultimately, what determines the appropriate scale to be used is the particular question being explored.

There are two important aspects of scale that we must take into consideration when designing our studies: the extent of the study area and the resolution at which we sample. Both aspects are correlated, and they set the natural boundaries of the scales we could use. A good way to visualize such considerations is through spatial-temporal domain diagrams, in which the hierarchical relationships of different scales for a given process can be represented (see, for instance, Delcourt et al. 1983). A few general patterns emerge from such exercises; for instance, studies at larger spatial scales tend to deal with greater variability and require longer temporal frames. We can use these as guidelines in choosing the scales of our studies.

Perhaps because of the overwhelming complexity of ecological systems, most studies focus on short-term and local dynamics. Yet, these are systems characterized by great variation in time and space. Understanding the causes and structure of these variations is essential for our ability to predict the future and our attempts to regulate these dynamics. To illustrate, until recently a considerable fraction of population modeling evaluated the variation of vital rates from data collected at small spatial scales and spanning few years with little regard for the full extent of environmental variation in time and space. This means changes in vital rates and population sizes cannot be linked to environmental changes or management regimes, effectively limiting our ability to understand and attempt to control their influence on populations.

A clear comprehension of temporal and spatial change as a function of variation in ecological drivers will be fundamental for sustainable resource use, management, and conservation. This type of approach delivers more mechanistic and predictive information to help our understanding of ecological dynamics. Studying the demographic effect of environmental drivers will allow us to project population changes due to novel conditions caused by human actions.

Practical insights

Fire is a major ecological disturbance widely spread throughout the world (Bond and Keeley 2005). The dramatic events in the grasslands and woody communities of Argentina, Australia, United States, Indonesia, the Pantanal, and the Amazonian basin in Brazil – and even the Arctic – during 2020 are evidence of the growing importance of this phenomenon.

Fire strongly affects ecosystems, communities, and populations. Extreme temperatures and severe droughts, caused by human-driven environmental change, have altered fire characteristics around the world, increasing both fire severity and incidence (Bowman et al. 2009). Wise use of fire suppression and prescribed fires can help to maintain disturbance patterns that are essential to many ecological processes, while minimizing negative economic and social impacts. Continually improving our understanding of fire ecology – which involves updating our knowledge under changing conditions – is imperative to design better management practices.

The regional spatial structure determines the availability of suitable habitats, dispersal distances, and levels of population and habitat isolation. The influence of patch dynamics is determined by correlations among patches affecting regional habitat heterogeneity and metapopulation persistence. For example, spatial proximity affects the spread of diseases and fire.

In previous chapters, we demonstrated the advantages of recognizing the nature of our sampling schemes and, when relevant, evaluating the effects of random factors in our models. In Chapters V and VIII, we described the association of number of fruits of *Hypericum cumulicola* with plant standing height. We also recognized there was variation in the number of fruits per plant among populations. Now, we want to evaluate the relevance of two more variables to explain variation in fecundity for this species. Time-since-fire (TSF) affects, among other things, nutrient and water availability, and the abundance of predators and competitors, potentially influencing plant attributes (such as size) and resources that influence reproduction. We also evaluate the covariance (see Chapter XI and below) among populations. For this purpose, we use a matrix of distance between populations, since we expect populations that are closer to each other to be environmentally more similar. A map of the Florida rosemary scrub at Archbold Biological Station (Figure 12.1) reveals the distribution of suitable habitats and the incidence of *H. cumulicola* within them.

Model building

We use a model selection approach to assess the relative importance of fire and relative population location to explain variation in fruit production of *Hypericum cumulicola*. We use information obtained from 14 populations to build our model set (the full annotated code and sample datasets can be found on the website). In this case, we focus only on data collected in 1995. We are still able to evaluate temporal variation, however, by using a design that substitutes elapsed time with spatial variation in time-since-fire. This is a suboptimal approach since it confuses both types of variation. Yet, because it allows the shortening of study times, it is fairly common in ecology, and thus it is worth exploring. An analysis of demographic effects of time-since-fire on this species using 22 years of longitudinal data is in Quintana-Ascencio et al. (2018 and 2019).

Here, we estimate time-since-fire as the difference between 1995 and the year of the last known prescribed fire for each population. We also use the logarithm of height as a predictor to help account for individual variation. We scale the data as standard normal deviates to facilitate calculations and to minimize the correlation between coefficients due to the interaction of time-since-fire and plant height. As mentioned in the previous section, we require the distances between the 14 populations to calculate their correlated influence as random effects.

When comparing mixed models, Zuur et al. (2009) recommend starting by evaluating different configurations of random effects. To do this, they suggest building models that use the same fixed effect combination (ideally including all possible variables). After identifying the most likely configuration of random effects, we move to evaluating different combinations for the chosen fixed effects.

For our data, the models for the first step include, as fixed effects, plant height (logarithmically transformed), time-since-fire, and a two-way interaction between the two. We propose three options for the random configuration: (i) no random effects, (ii) random effects on the intercept given the population, and (iii) correlated random effects among populations. These first two models should look familiar. The simplified versions of these models are in Boxes 12.1-12.3.

Box 12.1: Simplified implementation in *R* and *Stan* for Model 12.1 with no random effects

```
// likelihood
    ln (fruits) ~ normal(mu,sigma)

// model specification
    for (i in 1:N){
    mu[i] = a + b*ln(height) + c*tsf + cc*tsf*ln(height)}

// prior of intercept
    a ~ normal(0,50),
```

```
// prior of coefficient for height
    b ~ normal(0,1),

// prior of coefficient for TSF
    c ~ normal(0,1),

// prior of coefficient of interaction
    cc ~ normal(0,1),

// prior for variance
    sigma ~ normal(0,1)
```

Box 12.2: Simplified implementation in *R* and *Stan* for Model 12.2 with random intercepts

```
// likelihood
    ln (fruits) ~ normal(mu,sigma)

// model specification
    for (i in 1:N){
    mu[i]  =  a  +  p[pj]  +  b*ln(height)  +  c*tsf  +
    cc*tsf*ln(height)}

// prior of intercept
    a ~ normal(0,50),

// prior of coefficient for height
    b ~ normal(0,1),

// prior of coefficient for TSF
    c ~ normal(0,1),

// prior of coefficient of interaction
    Cc ~ normal(0,1)

// prior of random effects per population
    p[pj] ~ normal(0,sigmap)

// prior of random population variance
    sigmap ~ dcauchy(0,1),

// prior of variance
    sigma ~ dcauchy(0,1)
```

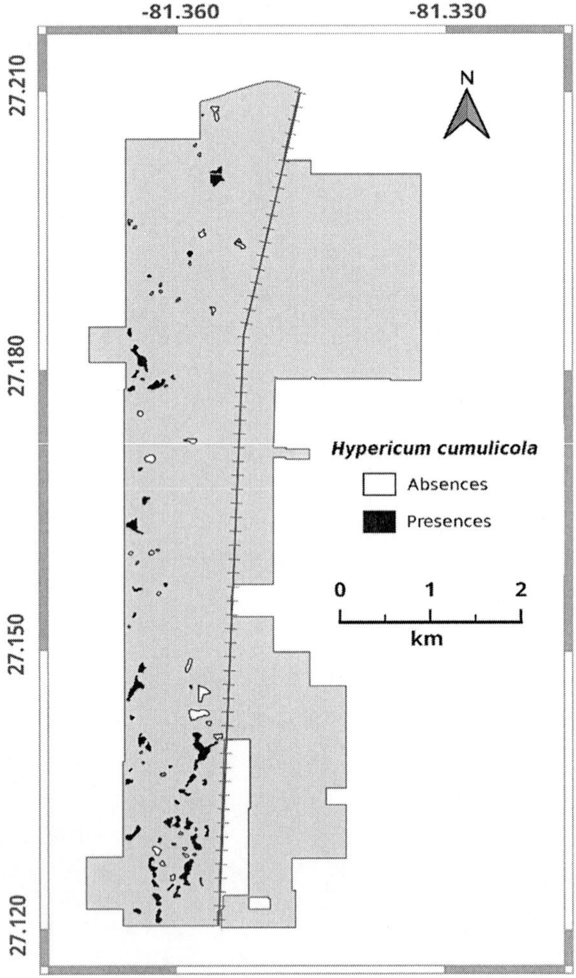

Figure 12.1: Florida rosemary scrub patches and *Hypericum cumulicola* occurrence in this habitat within Archbold Biological Station. This species only occurs in a fraction of possible suitable patches at the Station. Except for a large area at the upper western side and smaller areas to the southwest and southeast of the Station (not shown) nearby Florida scrub patches beyond the Station boundary were transformed into humanized habitats (orange groves, dwellings, and roads) and are no longer suitable

For the third model (Model 12.3), we use the distances (in km) between each population. The geographic distance matrix is displayed in Figure 12.2.

	1	29	32	42	50	57	59	62	67	87	88	91	93
1	0.0												
29	4.2	0.0											
32	4.2	0.2	0.0										
42	6.0	1.7	1.4	0.0									
50	6.8	2.5	2.2	0.6	0.0								
57	7.0	2.8	2.5	0.9	0.3	0.0							
59	7.1	2.8	2.6	0.9	0.3	0.1	0.0						
62	7.3	3.0	2.8	1.2	0.6	0.3	0.2	0.0					
67	7.6	3.4	3.1	1.5	0.8	0.6	0.5	0.3	0.0				
87	8.4	4.1	3.9	2.3	1.6	1.3	1.3	1.0	0.6	0.0			
88	8.3	4.1	3.8	2.1	1.5	1.3	1.2	0.9	0.5	0.3	0.0		
91	8.8	4.4	4.2	2.5	1.9	1.7	1.6	1.4	1.0	0.7	0.5	0.0	
93	8.8	4.5	4.2	2.5	1.9	1.7	1.6	1.4	1.0	0.5	0.4	0.1	0.0
103	3.4	0.6	1.0	2.5	3.3	3.6	3.6	3.9	4.2	4.9	4.9	5.3	5.3

Figure 12.2: Distance matrix among studied populations. The distance (km) between each pair is indicated below the diagonal. The darkness above the diagonal increases as distance decreases (0, 0.1-1, 1-2, 2-3, 3-4 and >4 km)

For example, population 29 is approximately 1.66 km from population 42. Notice that the diagonal is all zeros, and that the matrix is symmetric around the diagonal. These data will allow us to estimate varying intercepts for each population that account for non-independence in fruit production as a function of distance between populations. The first lines of the models are familiar. The *p[pj]* will be the varying intercepts in this case. We also include ordinary coefficients for the other variables. The highlight in this model is the hypothesized correlated effect of the populations, in this model defined by a covariance matrix ***K*** as the multivariate prior for these effects (after: McElreath 2016).

$$\gamma \sim MVNormal([0, \ldots .0], K) \qquad \text{(Equation 12.1)}$$

$$K_{ij} = \eta^2 \exp\left(-\rho^2 D_{ij}^2\right) + \delta_{ij}\sigma^2 \qquad \text{(Equation 12.2)}$$

The first equation (Equation 12.1) is a fourteen-dimensional Gaussian prior for the covariant effect of the populations. The vector of means for this effect includes only zeros because the grand mean is "*a*" in the model, making the population effects deviate from the mean expectation. The second equation (Equation 12.2) is their covariance matrix. Covariance measures the amount of variation when random variables vary together. Covariance can be zero, negative or positive. A positive value indicates that they tend to vary in the same direction, and a negative value implies they vary in opposite directions. Zero covariance means they vary with complete independence. Covariance is defined by:

$$Cov(XY) = \frac{\Sigma(x_i - \bar{x})(y_i - \bar{y})}{N-1} \qquad \text{(Equation 12.3)}$$

Also, this is the mean value of the product of the deviations of two random variables from their respective means.

Equation 12.2 uses three parameters, η, ρ, and σ to model how the covariance among populations changes with distance. The part $\exp\left(-\rho^2 D_{ij}^2\right)$, where D is distance, indicates that covariance between populations declines exponentially with the square of the distance. The parameter ρ determines the rate of decline. The last two pieces are η^2, the maximum covariance between two populations i and j, and $\delta_{ij}\sigma^2$, the extra covariance beyond η^2, when $i = j$. The model computes the posterior distribution of η, ρ, and σ, and needs priors for them.

Box 12.3: Simplified implementation in *R* and *Stan* for Model 12.3 with random spatially correlated population effects:

```
// likelihood
    ln (fruits) ~ normal(mu,sigma)

// model specification
    for (i in 1:N){
    mu[i]  =  a  +  p[pj]  +  b*ln(height)  +  c*tsf  +
    cc*tsf*ln(height)}

// prior of effects per population
      p[pj] ~ GPL2(Dmat,etasq,rhosq,sigmap),

// prior of intercept
```

```
    a ~ normal(0,50),

// prior of coefficient for height
    b ~ normal(0,1),

// prior of coefficient for TSF
    c ~ normal(0,1),

// prior of coefficient of interaction
    Cc ~ normal(0,1),

// prior of η
    etasq ~ cauchy(0,1),

// prior of ρ
    rhosq ~ dcauchy(0,1),

// prior of σ
    sigmap ~ exponential(1),

// prior of variance
    sigma ~ dcauchy(0,1)
```

The results obtained from running three MCMC chains with 10000 sampling iterations for Model 12.1 are in Table 12.1, those of Model 12.2 are in Table 12.2 and the ones for Model 12.3 are in Table 12.3.

Table 12.1: Summary of the parameters in Model 12.1, with no random effects. Mean and standard deviation of each parameter and their credibility intervals

Coefficient	**Mean**	**Std Dev**	**lower CI**	**upper CI**	**Rhat**
a, intercept	4.21	0.03	4.15	4.26	1.00
b, height	1.08	0.03	1.02	1.13	1.00
C, TSF	-0.06	0.03	-0.11	-0.01	1.00
*cc, Interaction height*TSF*	-0.07	0.03	-0.12	-0.09	1.00
Sigma	0.75	0.02	0.72	0.76	1.00

Table 12.2: Summary of the parameters in Model 12.2, with random effects due to population. Mean and standard deviation of each parameter and their credibility intervals

Coefficient	**Mean**	**Std Dev**	**lower CI**	**upper CI**	**Rhat**
a, intercept	4.21	0.09	4.02	4.40	1.00
B, height	1.12	0.03	1.07	1.18	1.00
c, TSF	-0.06	0.09	-0.25	0.12	1.00
*cc, interaction height*TSF*	-0.07	0.03	-0.13	-0.02	1.00
Sigma(p), random effects	0.33	0.08	0.20	0.53	1.00
Sigma	0.71	0.02	0.67	0.72	1.00

Table 12.3: Summary of the parameters in Model 12.3, with correlated random effects among populations. Mean and standard deviation of each parameter and their credibility intervals

Coefficient	**Mean**	**Std Dev**	**lower CI**	**upper CI**	**Rhat**
a, intercept	4.14	0.36	3.27	4.76	1.00
b, height	1.12	0.03	1.07	1.18	1.00
c, TSF	-0.04	0.11	-0.26	0.18	1.00
*cc, height * TSF*	-0.07	0.03	-0.13	-0.02	1.00
etasq	0.24	0.53	0.003	1.38	1.00
rhosq	5.67	76.59	0.004	27.84	1.00
Sigmap	0.14	0.09	0.039	0.37	1.00
Sigma	0.71	0.02	0.67	0.74	1.00

The model comparison in Table 12.4 indicates that, since Models 12.2 and 12.3 have more weight, there is clear evidence of differences among populations. However, there is a small change in weight between the WAIC of Models 12.2 and 12.3 (0.53 vs. 0.47). In this case, adding the spatial relationships among populations does not appear to provide much additional information. A weak spatial effect is consistent with the high population genetic differentiation and limited interpopulation gene flow documented by Dolan et al. (1999). On the other hand, critical habitat attributes are more similar between populations in nearby patches than between distant ones. Fire ignition and spread and relative elevation, a local indicator of water availability, are spatially correlated in this ecosystem (Quintana-Ascencio et al. 2018). What do you think? Is the additional complexity in Model 12.3 with information on spatial correlations among populations warranted? We should explore the whole variation associated with the fixed and random factors before deciding the inference from our models.

Table 12.4: Summary of model comparison using the WAIC among Models 12.1-12.3

Model	WAIC	dWAIC	weight
Random intercept (M-12.2)	1740.1	0.0	0.53
Random spatial (M-12.3)	1740.3	0.3	0.47
No random effects (M-12.1)	1823.9	83.8	<0.001

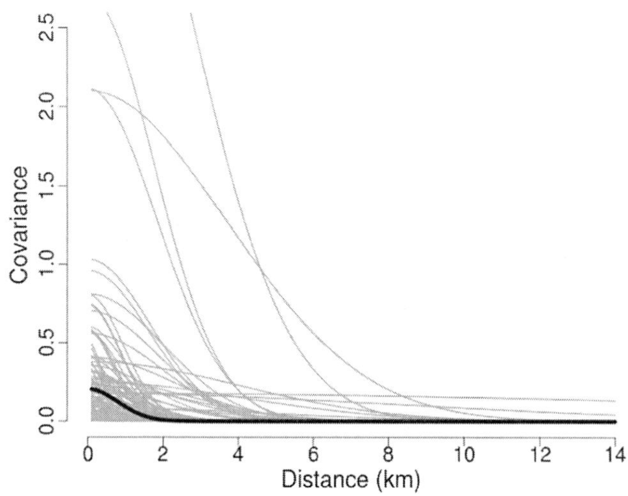

Figure 12.3: Posterior distribution of the spatial covariance between a pair of populations. The black line displays the posterior median. The gray lines show 100 realizations sampled from the joint posterior distribution of η, ρ

What about the correlations in individual fruit production among populations? We present the correlations in fruit production among populations in Figure 12.4. We start noticing that predicted correlations in fruit production per population using Model 12.3 are generally consistent with the distances among populations; for example, there are high correlations among populations 50, 57, 59 and 62 that are congregated in the center of the Archbold Biological Station, and between 87, 88, 90 and 91 which are close to each other near the south section of the station. Due to their proximity, they share relative elevation and time-since-fire, attributes that influence seed production (Quintana-Ascencio et al. 2019). Next, in Figure 12.5, we observe that the random effect structure of the model changes the inference on the fixed factors and on rho.

	1	29	32	42	50	57	59	62	67	87	88	91	93
1	1.0												
29	0.0	1.0											
32	0.0	0.9	1.0										
42	0.0	0.1	0.7	1.0									
50	0.0	0.0	0.0	0.2	1.0								
57	0.0	0.0	0.0	0.5	0.9	1.0							
59	0.0	0.0	0.0	0.5	0.9	0.9	1.0						
62	0.0	0.0	0.0	0.3	0.7	0.9	0.9	1.0					
67	0.0	0.0	0.0	0.2	0.5	0.7	0.8	0.9	1.0				
87	0.0	0.0	0.0	0.0	0.1	0.2	0.3	0.4	0.7	1.0			
88	0.0	0.0	0.0	0.0	0.0	0.2	0.3	0.3	0.4	0.9	1.0		
91	0.0	0.0	0.0	0.0	0.0	0.0	0.1	0.1	0.2	0.6	0.8	1.0	
93	0.0	0.0	0.0	0.0	0.0	0.0	0.1	0.1	0.2	0.8	0.8	0.9	1.0
103	0.0	0.7	0.3	0.0	0.0	0.0	0.0	0.0	0.0	0.0	0.0	0.0	0.0

Figure 12.4: Fruit production correlation matrix among studied populations. This matrix is symmetrical. The correlation coefficient between each pair is indicated below the diagonal. The darkness above the diagonal increases as correlation increases (<0.1, >0.1-0.39, 0.40-0.59, 0.60-0.89, >0.90)

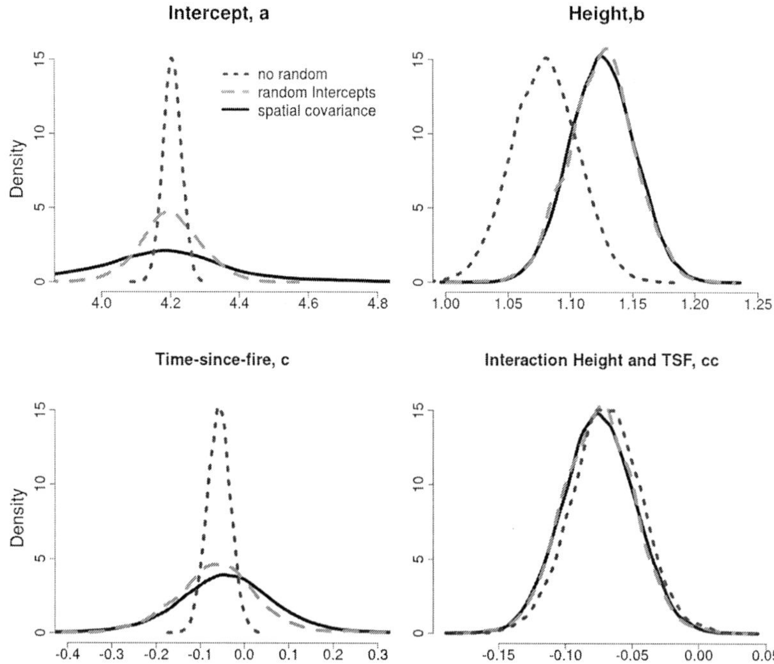

Figure 12.5: Posterior distribution of the coefficients of the fixed variables (a, intercept; b, height; c, TSF; and cc, the interaction between TSF and height) for models with a different structure of the random effects: (1) no random effects, (2) random intercepts, and (3) spatial covariant random effects

We now move to the second step recommended by Zuur and colleagues (2009): evaluating fixed effects. To do this, we create models that vary the fixed effects while maintaining the random effects as constant (using the formulation selected in the previous step). Arguably, we are more interested in the fixed factors. Notice that the inclusion of random effects increases the uncertainty of *a*, the intercept, and coefficient *c*, the one associated with time-since-fire.

We have already evaluated the saturated model (Model 12.3) and now we generate Model 12.4, with random correlated spatial effects by population but without interactions between height and time-since-fire, and Model 12.5, that does not evaluate the effect of time-since-fire, only the effect of plant height. The simplified implementations in *R* and *Stan* for Model 12.4 and Model 12.5 are in Box 12.4 and Box 12.5, respectively.

Box 12.4: Simplified implementation in *R* and *Stan* for Model 12.4 with random spatially correlated population effects, but no effect of the interaction between time-since-fire and height

```
// likelihood
    ln (fruits) ~ normal(mu,sigma)

// model specification
    for (i in 1:N){
    mu[i] = a + p[pj] + b*ln(height) + c*tsf}

// prior of effects per population
      p[pj] ~ GPL2(Dmat,etasq,rhosq,sigmap),

// prior of intercept
    a ~ normal(0,10),

// prior of coefficient for height
    b ~ normal(0,10),

// prior of coefficient for TSF
    c ~ normal(0,1),

// prior of η
    etasq ~ cauchy(0,1),

// prior of ρ
    rhosq ~ dcauchy(0,1),

// prior of σ
    sigmap ~ exponential(1),

// prior of variance
    sigma ~ dcauchy(0,1)
```

Box 12.5: The simplified implementation in *R* and *Stan* for Model 10.5 with random spatially correlated population effects but no effect of time-since-fire

```
// likelihood
    ln (fruits) ~ normal(mu,sigma)

// model specification
    for (i in 1:N){
    mu[i] = a + p[pj] + b*ln(height)}
```

```
// prior of effects per population
       p[pj] ~ GPL2(Dmat,etasq,rhosq,sigmap),

// prior of intercept
    a ~ normal(0,10),

// prior of coefficient for height
    b ~ normal(0,10),

// prior of η
    etasq ~ cauchy(0,1),

// prior of ρ
    rhosq ~ dcauchy(0,1),

// prior of σ
    sigmap ~ exponential(1),

// prior of variance
    sigma ~ dcauchy(0,1)
```

The results obtained from running three MCMC chains with 10000 sampling iterations for Model 12.4 are in Table 12.5 and those for Model 12.5 are in Table 12.6.

The comparison between all the models constructed so far in this chapter – presented in Table 12.7 – provides evidence of differences among populations and for the clear variation in number of fruits due to height, time-since-fire, and their interaction. Notice the apparently much larger uncertainty in the posterior distribution of ρ^2 (*rhosq*) in the more complex models (Table 12.3, Tables 12.5 and 12.6). These estimates are biased to the left but include a few large values, producing higher variances and larger means. The median values of these distributions are much lower and consistent (0.56, 0,53 and 0.52 respectively for the three models).

Table 12.5: Summary of the parameters in Model 12.4. Mean and standard deviation of each parameter and their credibility intervals

Coefficient	Mean	Std Dev	lower CI	upper CI	Rhat
a, intercept	4.14	0.32	3.44	4.76	1.00
b, height	1.12	0.03	1.07	1.18	1.00
c, TSF	-0.05	0.11	-0.27	0.18	1.00
etasq	0.23	0.55	0.002	1.37	1.00
rhosq	93.92	9094.81	0.005	22.20	1.00
Sigma (p)	0.14	0.09	0.04	0.37	1.00
Sigma	0.71	0.02	0.68	0.75	1.00

Table 12.6: Summary of the parameters in Model 12.5. Mean and standard deviation of each parameter and their credibility intervals

Coefficient	Mean	Std Dev	lower CI	upper CI	Rhat
a, intercept	4.10	0.32	3.34	4.2	1.00
b, height	1.12	0.03	1.07	1.18	1.00
etasq	0.23	0.68	0.003	1.31	1.00
rhosq	4.90	48.30	0.004	25.36	1.00
Sigma (p)	0.12	0.07	0.03	0.31	1.00
Sigma	0.71	0.02	0.68	0.75	1.00

Table 12.7: Summary of model comparison using the WAIC among models

Model	WAIC	dWAIC	weight
Random intercepts; 12.2	1740.1	0.0	0.51
Random spatial; 12.3	1740.3	0.3	0.45
Random spatial no fire; 12.5	1746.1	6.1	0.02
Random spatial no interaction; 12.4	1746.4	6.4	0.02
No random effects; 12.1	1823.9	83.8	<0.001

How do we interpret this information? Consider that the management practice of burning by spatial units creates a strong correlation among populations in relation with fire. This "collinearity" between spatial distance and time-since-fire among populations underscores the correlated random effect of populations on seed production and may create a certain redundance between these variables increasing the estimate of their uncertainty in those models.

Final considerations

It is critical to understand the ecological effects of environmental drivers and their spatial structure for wise management and sustainable use of resources. Mitigation of growing human impacts on many species will require obtaining information based on data collected with efficient and robust sampling designs. Evaluating these data using models of increasing complexity including the interacting effects of many variables to project meaningful scenarios will allow us to assess alternative scenarios for management and conservation.

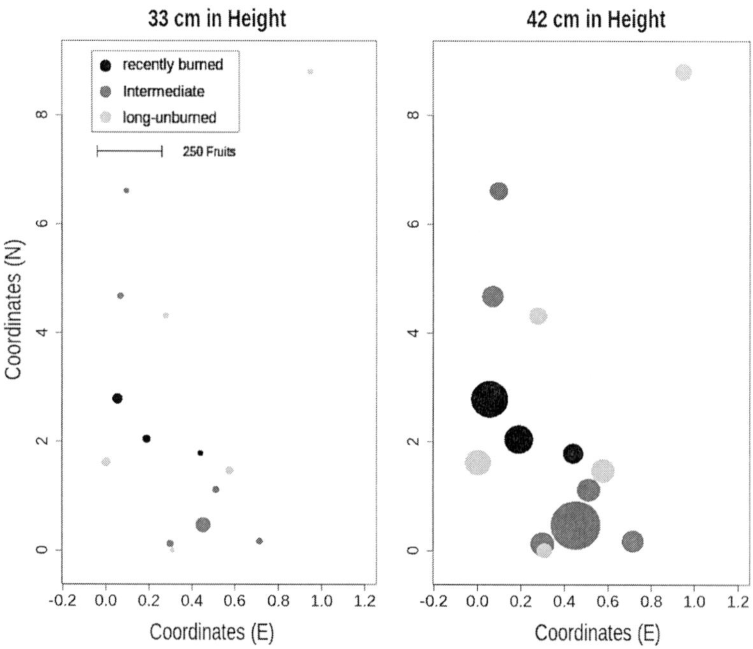

Figure 12.6: Plot of the spatial location of studied Florida rosemary scrub patches. The size of the spheres indicates the relative predicted (after Model 12.3) average number of fruits for plants of 33 and 42 cm in height per population. Shades of gray indicate different times-since-fire (2 years, 10-12 and >20 years) in 1995

In the example of this chapter, we provide evidence of the critical role of fire on the reproduction of *Hypericum cumulicola* (Figure 12.6). We document interesting but, based on the information from these models in this chapter, inconclusive effects of the correlated effects of distance among populations in different Florida rosemary scrub patches on individual fruit production. The relevance of dispersal, and the spatial distribution of individuals and populations at different scales, is documented elsewhere (Quintana-Ascencio et al. 1996, Dolan et al. 2008, Miller et al. 2012, Quintana-Ascencio et al. 2020, Coutts et al. 2021).

Chapter XIII

Integration: Population Dynamics in Natural and Humanized Environments

"Wilderness recovery, I firmly believe,
is the most important task of our generation."
– Reed Noss

Relevant notions

Statistical inference has frequently been limited to evaluating relatively simple hypotheses, often using the strength of association among variables as a measure of evidence. The identified parameters, however, can be combined in more elaborate models to evaluate complex phenomena. Causal models allow the organization of information to ask "what if?" questions that can help achieve deeper learning (Pearl and Mackenzie 2018).

The power of integrating statistical information is exemplified by integral projection models (IPMs), a powerful tool used in population ecology.

Integral projection models, and other structured demographic models, allow the evaluation and simulation of population dynamics in plausible scenarios. These models are comprised of statistical sub-models estimating vital rate variation – the types of models we have explored so far in the book. This approach has the advantage that the variation of different vital rates can be evaluated as a function of different drivers. It makes it possible to explicitly incorporate factors relevant to management, to explore life history, and to answer diverse ecological questions. Typically, IPMs use projections of survival and growth sub-models to build a state transition kernel and projections of fecundity sub-models to build a reproductive kernel. The projections from continuous functions are discretized and the resulting sub-matrices integrated into a matrix model.

Integral projection models require a clear understanding of the life history of the focal species to avoid spurious projections. They benefit from long-term data, which help to capture the vagaries of any population (Quintana-Ascencio et al. 2021). A clear diagram of the relationships among the life stages of the species, and their triggers, facilitates the visualization and construction of causal models (Pearl and Mackenzie 2018). These models are very flexible and have many applications in ecology, evolution, and conservation (Metcalf et al. 2013, Merow et al. 2014). A useful description of their implementation can be found in Ellner, Childs and Rees (2016).

-- --

🕶 Glasses On 🕶

Integral projection modeling involves estimating vital rate variation as a function of, at least, a continued state variable – most frequently size. In its simpler form, it requires the estimation of two kernels. The first kernel describes the transition of individuals among realizations of a continuous state variable using estimates of individual growth and survival. The second defines the change in the number of individuals in the population due to recruitment.

Every IPM consists of a single or several merged rectangular matrices to consider, for example, vegetative and adult transitions. They can contain several vectors to account, for example, for the germination of dormant seeds vs. the germination from recently produced seeds. They can also include scalars describing, for instance, long-term seed dormancy (see, for instance, Tye et al. 2016).

The mid-point rule approximation to the continuous form of these IPMs is constructed as a Goodman matrix (Goodman 1969) model with a specified number of discrete classes (*m*) describing the population. The responses of coupled equations are integrated over *L*, the minimum expected size, and U, the maximum expected size of width = *(U-L)/m*. It is recommended to evaluate different amounts of classes to avoid any bias but minimize computation problems. The discrete form of the IPM is described as (Ellner, Childs & Rees 2016, González et al. 2021):

$$K(z_{jk}, z_{ik}) = P_{(j,i)} + F_{(j,i)} \qquad \text{(Equation 13.1)}$$

$$P_{(j,i)} = \sigma_{(j,i)} \times \gamma_{(j,i)} \qquad \text{(Equation 13.2)}$$

$$F_{(j,i)} = \text{Seed}_{(j,i)} + \text{Yearling}_{(j,i)} \qquad \text{(Equation 13.3)}$$

$$n(z_{jk}, t+1) = h \sum_{i=1}^{m} K\left(z_{jk}, z_{ik}\right) n(z_{ik} t) \qquad \text{(Equation 13.4)}$$

where *K* is the continued form of the model composed of two kernels *P* (growth and survival) and *F* (fecundity). The continuous form of this model is discretized to build a matrix model with separate stages.

-- --

Practical insights

In this chapter, we assess an example of a phenomenon that is widespread in anthropogenic environments. Oddly, the creation or alteration of habitats by humans has favored the increase of populations of certain species in the margins of native habitats, even for some rare and endemic species. This may seem like a bit of good news to conservationists. However, it is critical to improve our understanding of these changes before we celebrate the apparent success and become complacent. It is possible that populations in these anthropic environments have different dynamics, affecting their long-term persistence and evolution, that may not be compatible with their natural habitats – analogous to the dynamics of wild plants in agricultural fields or animals in zoos.

Hypericum cumulicola is mostly limited to gaps between vegetation patches in Florida rosemary scrub, but also occurs in fire lanes and on the shoulders of dirt roads adjacent to its natural populations (Quintana Ascencio et al. 2007). Managers in natural sanctuaries containing populations of this species have occasionally observed large numbers of *H. cumulicola* and several other Florida gap specialists in fire lines and other areas disturbed by humans and wondered about the conservation implications for these rare and endangered species. To tackle this question, a collaboration between managers and researchers was initiated and data have been collected at the Lake Wales Ridge Forest in Central Florida since 1996. In this chapter, we use integrative projection models to compare the demography of *Hypericum cumulicola* in Florida scrub patches (thereafter scrub) with the demography in dirt roads (hereafter roads).

The original data were collected in 20, 1 × 2 m, permanent plots within two replications for each type of habitat (Quintana Ascencio et al. 2007), but here we only use a portion of these data to illustrate the integration of statistical analysis in causal models in the form of integrative projection models. We refer the reader to the classic article by Hurlbert (1984) on the problems of pseudoreplication when improperly inferring from field experiments where treatments are not replicated and statistically controlled. Proper replication, randomization, and interspersion are a safeguard against confounding effects, unexpected events, and preexisting gradients. We use the data in this example only for illustration. We concentrate on data of populations that were not subject to fire during the study years (1996-ongoing) to avoid confounding effects with this important driver of the demography of *Hypericum cumulicola*. The chosen Florida scrub was particularly open, including large gaps, and located in a relatively high elevation ridge, where implemented prescribed fires did not penetrate. The chosen road is a 5-7 m wide linear lane with little traffic.

The examination of the size and age distribution between populations in these examples of the two habitats indicates that the road population has more left-skewed size distributions with proportionally more small individuals and more individuals dying younger (Figure 13.1). These attributes are consistent with the contrasting evolutionary strategies of species in environments with different levels of disturbance.

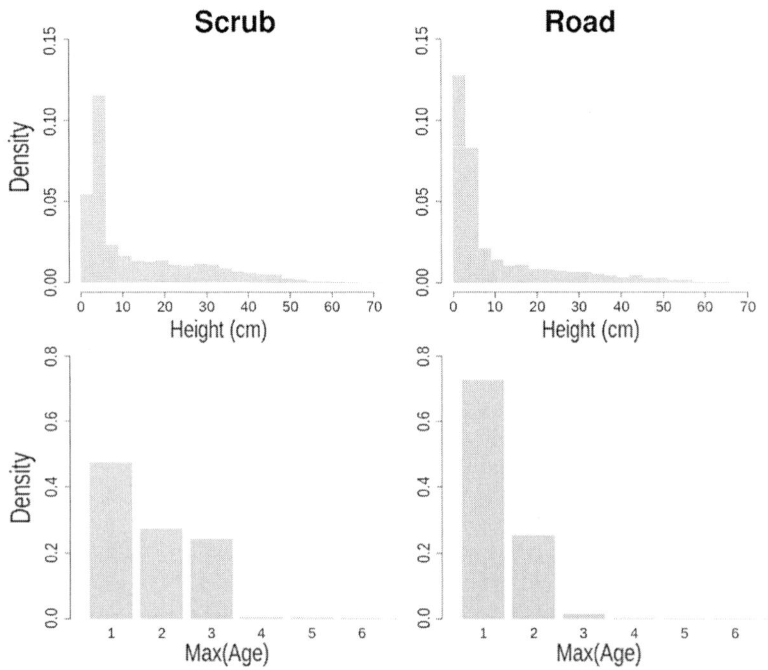

Figure 13.1: Histograms of standing height and maximum age of *Hypericum cumulicola* in populations at Florida scrub and road habitats in the Lake Wales Ridge Forest (1996-2017)

Model construction

We use the posteriors of models of *H. cumulicola* vital rates to build integral projection models simulating the dynamics of this species in scrub and road habitats. These models are used to obtain an estimate of the stochastic population growth rates (λ_s) in scrub and road using the equation (Caswell 2000):

$$\widehat{\log \lambda_s} = \frac{1}{T} \sum_{t=0}^{T-1} r_t \qquad \text{(Equation 13.5)}$$

where $\log \lambda_s$ is the logarithm of the stochastic lambda, N_t is the population size at time t, T is the total time interval and $r_t = \log((N_{t+1}) / N_t)$.

To build this type of population model, we should start by proposing a life history diagram based on census intervals, in our case defined by annual August censuses of this species (Figure 13.2).

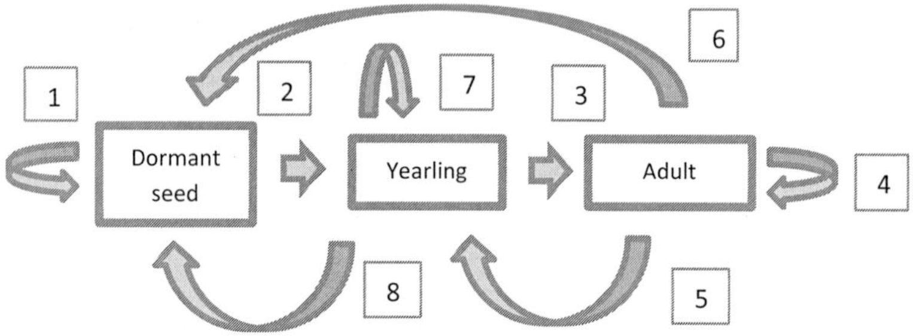

Figure 13.2: Diagram of the *Hypericum cumulicola* life history. The annual transitions include dormant seeds that can remain dormant (1) or germinate becoming yearlings (2) growing as adult plants (3). Adult plants can remain alive (4). Adults produce seeds that can germinate and become recruits during the same year (5) or remain as dormant seeds (6). Yearlings can also produce seeds that can germinate and become recruits during the year they are produced (7) or remain as dormant seeds (8). At any stage individuals can die. The transition among recruits and adults, and the production of seeds depend on size

We identify three main life history stages for this species: (1) dormant seeds, (2) yearlings, newly recruited plants reaching their first census year, and (3) adults, plants with more than one year of life (Quintana-Ascencio et al. 2020). Notice that we did not include recently produced seeds since they are "invisible" to the annual model proposed. By the annual census, recently produced surviving seeds either germinated and became yearlings or became dormant seeds. Dormant seeds were assumed as having a single state (discrete binomial probability values) while the variations of the vital rates of yearlings and adults were assumed to vary in a continuous fashion in response to their size (standing height) variation. Newly recruited plants behave differently compared to older plants (see Chapter III). We used the natural logarithm of individual standing height (size) as a continuous state variable because of its mathematical benefits.

Standing individuals at the census have either been recruited (as yearlings or dormant seeds) or survived and grown, and some of them reproduced. We model the variation in these four vital rates as a function of height (except for seeds), habitat (either scrub or road) and year using generalized linear mixed models (GLMM; Figure 13.3). Variation within a year was included as a random effect in all models when data permitted (α_{y_i}). We used binomial error distributions for survival (σ) and probability of reproduction (φ_0), and normal distributions for changes in the size of yearlings and adults (γ) and counts of reproductive structures (φ_1) because of the large number of fruits per plant (see Chapter IV). We used model selection based on WAIC information criteria (see Chapter VII) to justify the inclusion of habitat and plant size and the shape (linear or unimodal) of their functional relationship in models describing the variation of each vital rate (recruitment, survival, growth, probability of reproduction, and fecundity).

Seed dynamics (recruitment and survival) required estimates of probability of seed dormancy (s=0.65), probability of plant reproduction (φ_0), number of reproductive structures (φ_1), fruit set (φ_2=0.55), number of viable seeds per fruit (φ_3; average estimates from Chapter VII) proportion going dormant (δ=1-0.001), and seed survival through the census (φ_4=0.38).

$$SB(t+1) = SBs(t) + \int_L^U \varphi_0(x)\,\varphi_1(x)\varphi_2\varphi_3\varphi_4 s\delta dx$$

(Equation 13.6)

Recruitment of yearlings included estimates of the number of seeds in the seed bank ($SB(x,t)$), probability of reproduction (φ_0), number of reproductive structures (φ_1), fruit set (φ_2), number of viable seeds per fruit (φ_3), and germination probability ($\xi = 0.001$).

$$Yearlings(t+1) = SBs(t)\xi + \int_L^U \varphi_0(x)\,\varphi_1(x)\varphi_2\varphi_3 s\xi dx$$

(Equation 13.7)

Yearlings were assigned initial heights using observed normal probability distributions of logarithmic yearling height with mean and variances determined per habitat ($\bar{X}$ = ln (5.32 cm) $\pm$ 0.76 SE for scrub habitat and $\bar{X}$ = ln (5.08 cm) $\pm$ 0.80 SE for road habitat). The scripts for these models are similar to those developed in the previous chapters.

Modeling results

The averages and standard errors of the parameters estimated per model are in Table 13.1, and the plots of the model average predictions with 95% credibility interval envelopes are in Figures 13.3 to 13.6. There are many differences between populations in the scrub and on the road. We observe, for example, that the annual survival of small yearlings was higher in the scrub than on the road (approximately 0.6 and 0.5 respectively; Figure 13.3).

Table 13.1: Average coefficient estimates (with standard error in parentheses) for the most likely models in the set for the vital rates: growth (Gro), survival (Sur), probability of reproduction (Prep) and fecundity (Fec). The variables assessed were: h = height, h^2 = height^2, S = stage (adult as reference), R = habitat (scrub as reference), year as a random effect (Year) and error (Sigma), * indicates an interaction between variables. Not all possible interactions were found likely for these data

	Intercept	**h**	**h^2**	**S**	**R**	**H*Y**
Gro	1.67 (0.11)	0.55 (0.03)		0.25 (0.06)	0.56 (0.08)	-0.18 (0.04)
Sur	-5.31 (0.56)	4.08 (0.44)	-0.81 (0.09)	4.91 (0.61)	3.17 (0.50)	-2.76 (0.56)
Prep	-17.54 (0.93)	6.27 (0.31		-2.35 (1.03)	6.09 (1.09)	0.64 (0.36)
Fec	2.48 (1.68)	-2.59 (1.00)	0.86 (0.15)		-0.11 (0.06)	

Table 13.1: continuation

	h^2*Y	**h*R**	**h^2*R**	**Year**	**Sigma**
Gro				0.34 (0.06)	0.67 (0.02)
Sur	0.40 (0.12)	-4.31 (0.52)	0.86 (0.11)	0.98 (0.18)	
Prep		-2.02 (0.37)		1.03 (0.21)	
Fec					0.71 (0.02)

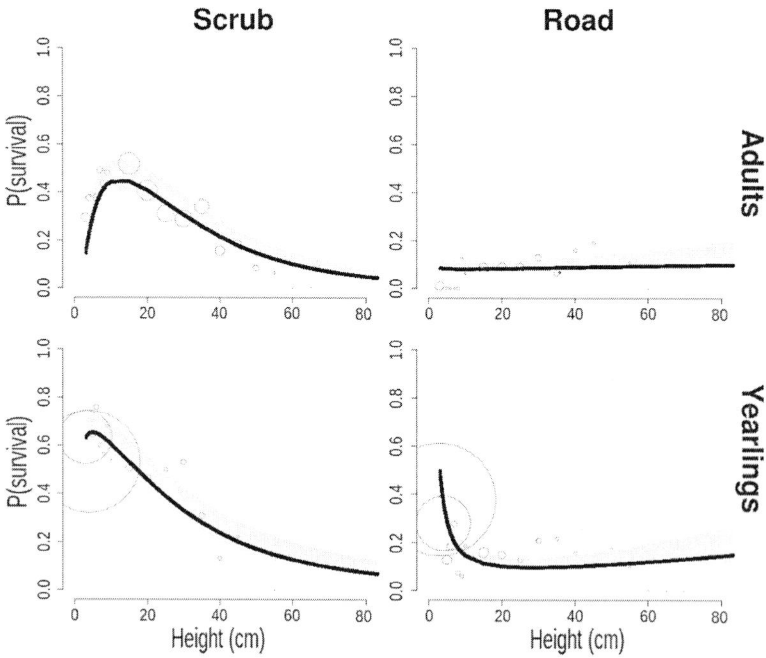

Figure 13.3: Plot of survival probability of *H. cumulicola* as a function of plant height for adult and yearling plants in scrub and road populations. The center of the circles indicates the average observed proportion of plants (binned at regular intervals) surviving. The size indicates the sample size.

The survival of yearlings decreased slowly with height in the scrub and rapidly on the road. Annual survival of adults in the scrub increased with plant height reaching a maximum (approximately 0.4) at 15 cm of height and then declined (Figure 13.3). Adult survival on the road was generally lower. Plants smaller than 15 cm (log = 2.7), the height when plants start reproducing, grew faster on the road than in the scrub (Figure 13.4). Plants on the road reached reproductive maturity at smaller heights (Figure 13.5). However, the fecundity of the plants on the road was lower (Figure 13.6)

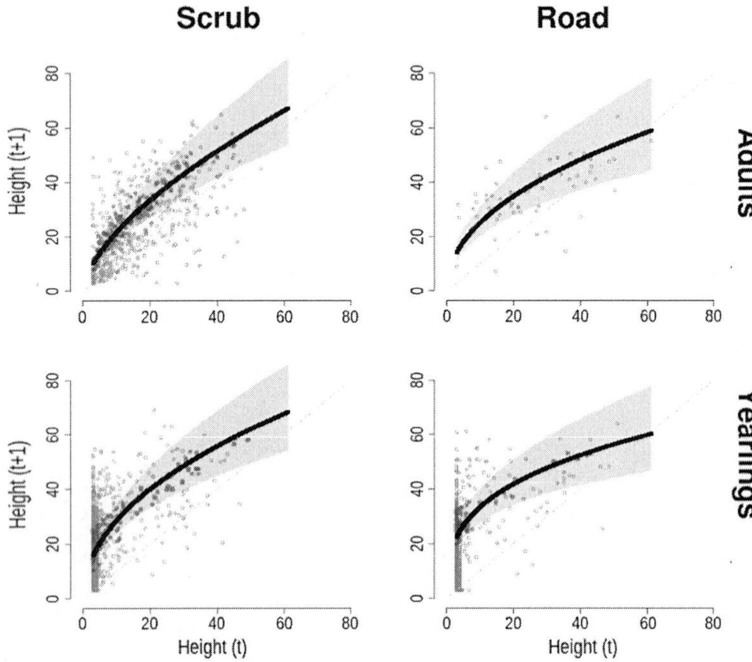

Figure 13.4: Plots of models of height (cm) of *Hypericum cumulicola* (in $t+1$) as a function of plant height (in t) by stage (yearlings or adults) and habitat (Florida scrub or road). Dots in plots indicate height for observed individuals per habitat. The discontinuous line in gray is the identity

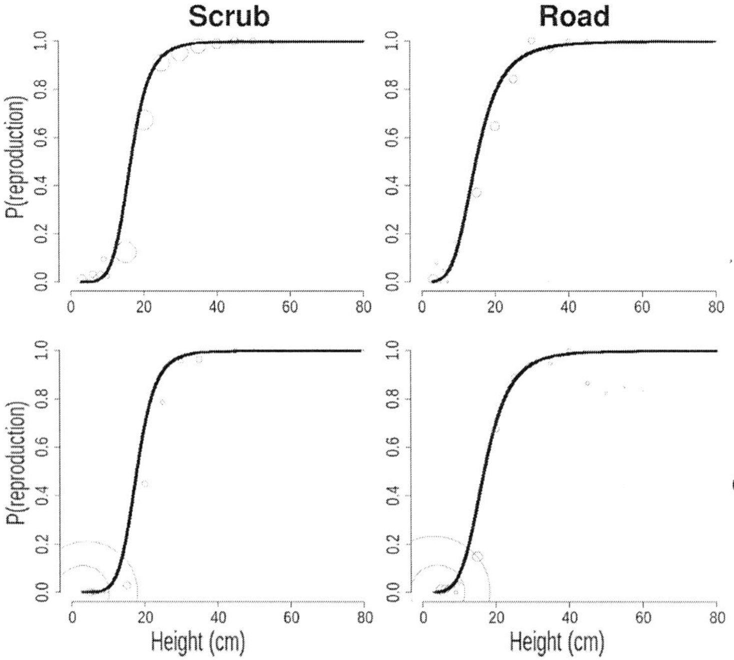

Figure 13.5: Plot of the reproduction probability of *Hypericum cumulicola* as a function of plant height (in cm) for adult and yearling plants in scrub and road populations. The center of the circles indicates the average observed proportion of plants (binned at regular intervals) surviving. The size of the circles indicates the relative sample size for the bin

We used the posterior predictions of the plausible models to estimate the responses of the different vital rates to state (yearling or adult), height, habitat (scrub or road), and year to build kernels of integral projection models for each habitat and year (Figure 13.7). The eigen-structure of these models quantifies estimates of the deterministic population growth rate (λ) for each habitat and year. We used matrix projection including random year effects for each vital rate to obtain the estimates of stochastic population growth per habitat and year.

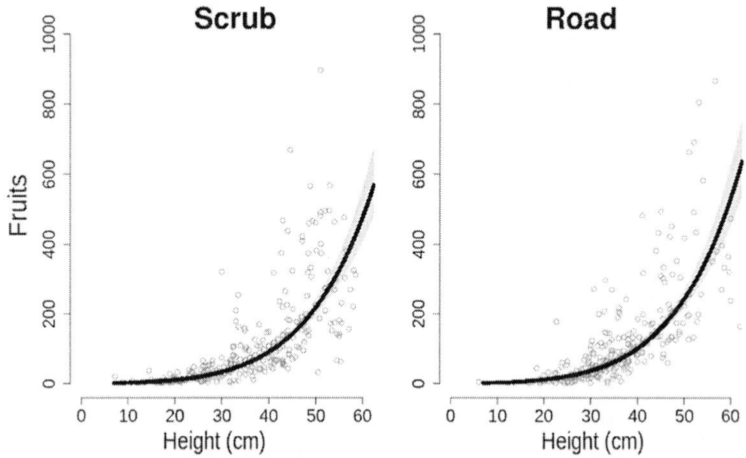

Figure 13.6: Plots of models of the number of fruits of *Hypericum cumulicola* as a function of plant height (cm) by habitat (Florida scrub or road). Dots in plots indicate height for observed individuals per habitat. We use projections of the model of adults for both stages because of the scarce number of fruits produced by yearlings

Figure 13.7 is a heat map of a matrix where lighter cells represent higher values. Columns are for time *t* stages and rows are for time *t+1* stages. Inside are all the transitions among them. This figure represents the kernels for each habitat obtained from the results of one iteration of each vital rate model. However, because all these models were constructed using MCMC sampling, we can construct multiple iterations of the kernel collectively representing the uncertainty of their configuration. Effectively, this gives us the advantage of being able to estimate the uncertainty around any derived parameter (such as the growth rate λ). In this figure each kernel has been separated into different segments. It helps to visualize each of the key transitions first described in Figure 13.2. Notice also how the upper left segment is not associated with a number. This is because this segment represents the impossible situation of a seed germinating into an adult. This is precluded by definition in the model, which is why the segment only contains zeros. On the other hand, the lower left segment represents the recruitment of new plants from the seed bank.

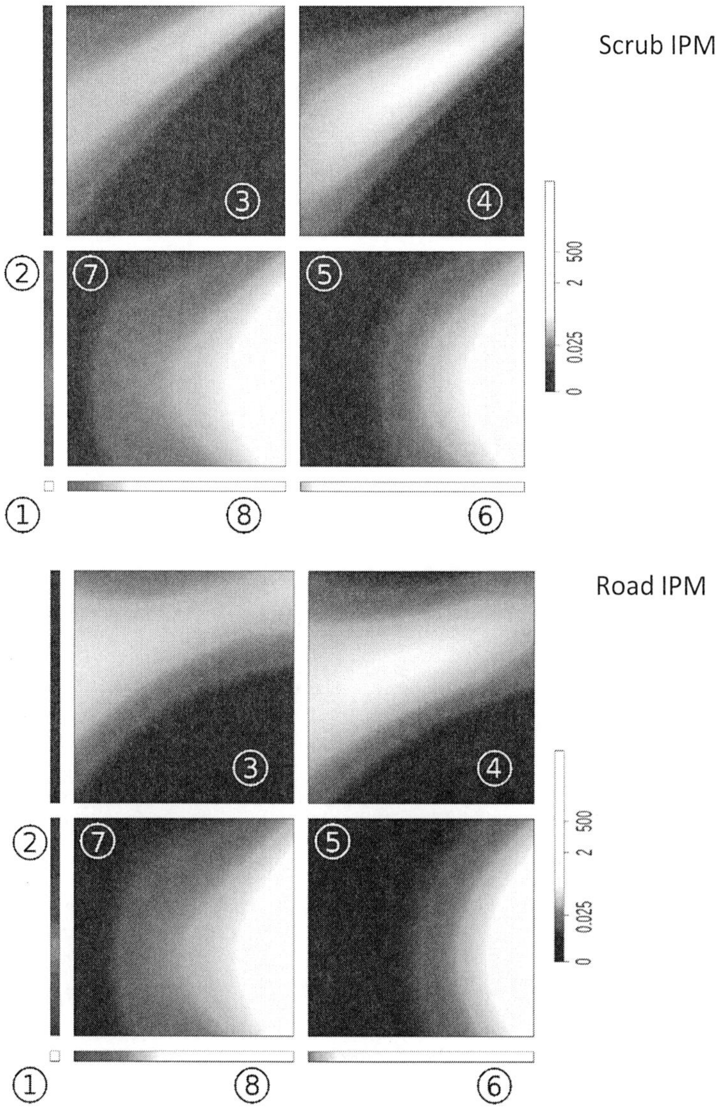

Figure 13.7: Examples of integral projection models of *Hypericum cumulicola* in the scrub (upper plot) and on the road (lower plot). Numbers inside the rings match those in Figure 13.2

Final considerations

Plants in the road habitat grew faster, started reproducing earlier and had a higher probability of reproduction, but they produced slightly fewer fruits for a given size, and died earlier than in their native habitat, the Florida rosemary scrub. Are the plants on the road potentially becoming weeds with a shorter life cycle and greater dependence for persistence on recruits and seed dormancy? You could argue that in an evolutionary sense this may be true.

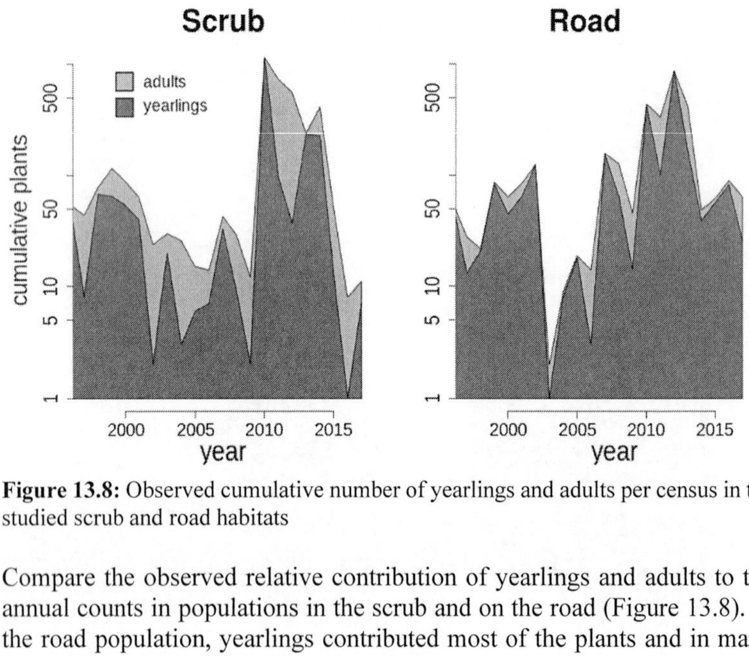

Figure 13.8: Observed cumulative number of yearlings and adults per census in the studied scrub and road habitats

Compare the observed relative contribution of yearlings and adults to the annual counts in populations in the scrub and on the road (Figure 13.8). In the road population, yearlings contributed most of the plants and in many years all the plants on the site.

The model projections correctly depicted this difference in relative numbers (lowest plots in the panel in Figure 13.9). The model predicts on average slow population declines for both populations (average lambda = 0.93 for the scrub and 0.86 for the road, respectively), a result consistent with previous projections for populations in the Florida scrub without fire (Quintana-Ascencio et al. 2019). Interestingly, the variance in the rate of population increase was larger in the road population than the scrub population. These results indicate that populations in the road habitat may be more prone to extinction and subject to selection for traits that are not compatible with the native habitat where plants respond to a gradient of time-since-fire.

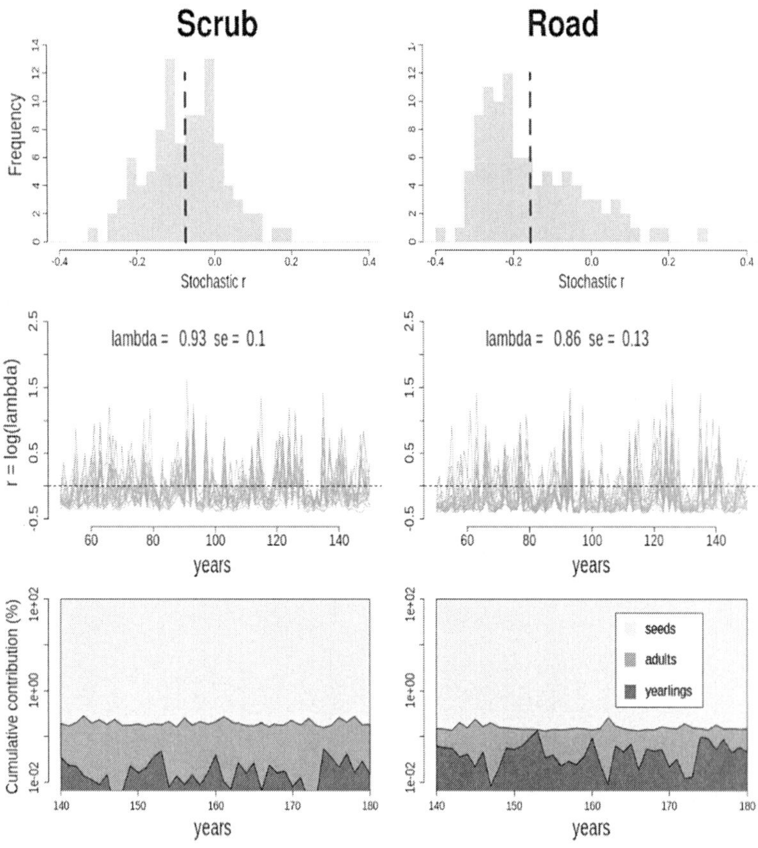

Figure 13.9: Upper plots: Frequency distribution of the projected stochastic finite rate of increase (r = log(*lambda*); n=100) per habitat. Middle plots: Examples (n=20) of sequences of projected r per year. Lower plots: Average predicted cumulative percentual contribution by seeds, yearlings and adults in scrub and road habitats. The year interval selected for the middle and lower plots was arbitrary

Reality, however, is more complicated. *Hypericum cumulicola* appears to have limited genetic variation for important ecological traits which could slow adaptive changes in novel habitats (Oakley 2013, 2015). The good news is that because of limited phenotypic differentiation combined with strong heterosis after crosses among populations there is also little risk of outbreeding depression in the native habitat (Oakley 2013, 2015).

CHAPTER XIV

MODELING HUMANIZED ENVIRONMENTS

"Change is coming whether you like or not."
– Greta Thunberg

Change is perhaps the most defining experience in our lives. In nature as in the science devoted to it, we are witnessing exceptional transformations that greatly affect the dynamics of natural systems and the way that we consider them. Never before have people experienced changes in their familiar landscapes with the magnitude seen by current generations. Natural systems are increasingly lost to agricultural, industrial, or urban developments, and those that remain are being degraded through pollution, global change, and destruction at a worrying pace. In his controversial book "The Unhabitable Earth: life after warming," David Wallace-Wells recently discussed in detail the concerning consequences of such degradation. Species extinction is now a threat even for common and previously widespread species. Many ecosystem services are at the point of collapse and essential resources are exhausted or poisoned. The social and economic consequences of this devastation are pervasive.

There is an urgent push to approach ecological studies with a pragmatic attitude aimed at understanding and mitigating the causes of these transformations. In the past couple of decades, ecological theory has shifted its focus to reflect this need. A field long concerned with describing the principles behind the organization of life is now confronted with the inescapable reality of a changing world. In this context, fundamental theory cannot be studied in isolation but must be integrated with applied theory. Issues that were previously sidelined or considered in isolation, such as seasonality, productivity, and disease dynamics, are now integral to ecological studies at every level (from individuals to landscapes). These issues are essential to understanding variation and unraveling the mechanisms of variation is essential in the age of climate change.

We are convinced that the transformation of our science is essential if we wish to contribute to the knowledge and management of these transformed systems and minimize their deterioration. We agree with the ideas expressed by Isabelle Stengers in her book "Another science is possible: a manifesto for slow science", that to tackle these and other urgent problems, science will need to be socially responsible. It is necessary to take an active role in identifying our biases when selecting the direction, design, and methods of our studies, as well as the possible consequences of the generated knowledge. It will also require a change in the approaches used to gain this knowledge, favoring those that recognize the complexity of these phenomena and the multiple scales of their causality and manifestation. It will be central to recognize and address the fast pace of their change. These are some of the challenges confronted by new generations of scholars.

Fortunately, current changes in ecological theory have been accompanied by an explosion in analytical resources, and, more importantly, a fundamental change in the way they are disseminated and evaluated. New developments in statistical analysis, mathematical modeling, and associated computer programs offer powerful and innovative platforms that can support transformative research. The robust sharing of code by supporting communities has changed the way information is used and improved. This book presents a glance at some of these developments. We hope our readers will continue exploring them and help with their dissemination and advancement.

Increased computational power, however, requires greater intellectual responsibility. As we start using more complicated analytical methods, there is an increased risk of losing the intuitive understanding of how underlying assumptions relate to the results. Even worse, in the frenzy for complexity, we might become blind to the said assumptions. These new methods can be very powerful predictive tools. However, to derive useful knowledge from our predictions, we must be purposeful in the development of our models. Our ambition with this book was to illustrate the types of questions that can help to guide the modeling process.

At the very least, we hope we were successful in depicting the complex nature of the kinds of systems we study in ecology. Even the simplest of questions can be approached in meaningful and interesting ways. We attempted to introduce some examples of current approaches used to describe and evaluate the attributes of these systems, hoping to illustrate how robust and powerful these tools can be. We are passionate about these topics because they are exciting and useful, and we hope our readers will continue exploring them.

BIBLIOGRAPHY

Abrahamson, W., Johnson, A., Layne, J., and P. Peroni. 1984. Vegetation of the Archbold Biological Station, Florida: An example of the southern Lake Wales ridge. Florida Scientist, 47, 209–250.

Aho, K., D. Derryberry and T. Peterson. 2014. Model selection for ecologists: the worldviews of AIC and BIC. Ecology 85: 631-636.

Beckmann, P. 1971. A history of π (Pi). St. Martin's Press, New York.

Benedict, R. 2005. Patterns of culture. Houghton Mifflin Harcourt. Boston, MA, USA.

Berkson, J. 1944. Application of the logistic function to bio-assay. Journal of the American Statistical Association 39: 357-365.

Betts, M.G. A.S. Hadley, D.W. Frey, S.J.K. Frey, D. Gannon, S.H. Harris, H. Kim, U.G. Kormann, K. Leimberger, K. Moriarty, J.M. Northrup, B. Phalan, J.S. Rousseau, T.D. Stokely, J.J. Valente, C. Wolf and D. Zárrate-Charry. 2021. When are hypotheses useful in ecology and evolution. Ecology and Evolution, https://doi.org/10.1002/ece3.7365.

Bond, W.J. and J.E. Keeley. 2005. Fire as a global 'herbivore': the ecology and evolution of flammable ecosystems. Trends in Ecology and Evolution, 20, 387-394.

Bond, W.J., Woodward, F.I. and G.F. Midgley. 2005. The global distribution of ecosystems in a world without fire. New Phytologist, 156, 525-538.

Bowman, D.M., Balch, J.K., Artaxo, P., Bond, W.J., Carlson, J.M., Cochrane, M.A., D'Antonio, C.M., DeFries, R.S., Doyle, J.C., Harrison, S.P. and F.H. Johnston. 2009. Fire in the earth system. Science, 324, 481-484.

Brudvig, L. and P.F. Quintana-Ascencio. 2003. Herbivory and post-grazing response in *Hypericum cumulicola*. Florida Scientist 66: 99-108.

Burgman, M. 2011. Remedies for a scientific disease. Bulletin of the British Ecological Society 42: 34-37.

Burnham, K.P. and D.R. Anderson. 2002. Model selection and multimodal inference: a practical information- theoretic approach. Springer.

Caswell, H. 2000. Matrix population models. Vol.1. Sunderland, MA, USA: Sinauer.

Choy, S.L., O'Leary, R. and K. Mengersen. 2009. Elicitation by design in ecology: using expert opinion to inform priors for Bayesian statistical models. Ecology, 90: 265-277.

Christman, S. P., and Judd. 1990. Notes on plants endemic to Florida Scrub. Florida W. S. Scientist, 53, 52–73.

Consul, P.C. 1989. Generalized Poisson Distributions: Properties and Applications. Marcel Dekker.

Coutts, S.R., Quintana-Ascencio P.F., Menges, E.S., Salguero-Gómez, R., and D. Z. Childs. 2021. Fine-scale spatial variation in fitness is comparable to disturbance-induced fluctuations in a fire-adapted species. Ecology 102: https://doi.org/10.1002/ecy.3287

Crone, E.E, M. Ellis, W.F. Morris, A. Stanley, T. Bell, P. Bierzychudek, J. Ehrlén. T. N. Kaye, T. M. Knight, P Lesica, G. Oostermeijer, P. F. Quintana-Ascencio, T. Ticktin, T. Valverde, J. L. Williams, D. F. Doak, R. Ganesan, K. McEachern, A. S. Thorpe and E. S. Menges. 2013. Ability of matrix models to explain the past and predict the future of plant populations. Conservation Biology, 27: 968-978.

Crawley, M.J. 2007. The R Book. Wiley.

Delcourt, P. A., and H. R. Delcourt. 1983. Late-Quaternary vegetational dynamics and community stability reconsidered. Quaternary Research 19: 265-271.

Dietze, M.C. 2017. Ecological forecasting. Princeton University Press. Princeton. New Jersey.

Dolan, R.W., Yahr, R., Menges, E.S. and M.D. Halfhill. 1999. Conservation implications of genetic variation in three rare species endemic to Florida rosemary scrub. American Journal of Botany, 86: 1556-1562.

Dolan, R. W., Quintana-Ascencio, P. F. and E. S Menges. 2008. Genetic change following fire in a population of a seed-banking perennial plant. Oecologia, 158:355-360.

Dorazio, R.M. 2016. Bayesian data analysis in population ecology: motivations, methods, and benefits. Population ecology, 58: 31-44.

Easterling, M.R., Ellner, S.P. and P.M. Dixon. 2000. Size-specific sensitivity: applying a new structured population model. Ecology, 81, 694–708.

Elfving, D. C., and I. Schechter. 1993. Fruit count, fruit weight, and yield relationships in delicious' apple trees on nine Rootstock." HortScience 28: 793-795.

Ehrlén, J. and W.F Morris. 2015. Prediction changes in the distribution and abundance of species under environmental change. Ecology Letters, 18, 303-314.

Ehrlén, J., Morris, W.F., von Euler T. and J.P. Dahlgren. 2016. Advancing environmental explicit structured population models of plants. Journal of Ecology, 104, 292–305.

Elderd, B. and T.E.X. Miller. 2016. Quantifying demographic uncertainty: Bayesian methods for integral projection models. Ecological Monographs 86: 125-144.

Ellison A.M. 2004. Bayesian inference in ecology. Ecology Letters 7: 509-520.

Ellner, S.P., Childs, D.Z. and M. Rees. 2016. Data-driven modelling of structured populations: a practical guide to the Integral Projection Model. Springer Publishing, Switzerland.

Ellner, S.P. and M. Rees. 2006. Integral projection models for species with complex demography. American Naturalist, 167, 410–428.

Evans, M.E., Menges, E.S. and D.R. Gordon. 2003. Reproductive biology of three sympatric endangered plants endemic to Florida scrub. Biological Conservation, 111: 235-246.

Famoye, F. and K.P. Singh. 2006. Zero-inflated Poisson regression model with an application to domestic violence data. Journal of Data Science 4: 117-130.

Forshey, C. G., and D. C. Elfving. 1989. The relationship between vegetative growth and fruiting in apple trees. Horticultural reviews 11: 229-287.

Gelman, A., 2011. Induction and deduction in Bayesian data analysis. Rationality, Markets and Morals, 2: 67-78.

Gelman, A., J. Hwang and A. Vehtari. 2014. Understanding predictive information criteria for Bayesian models. Statistics and Computing 24: 997-1016.

González, E. J., R Salguero-Gómez, D. Z. Childs, and P.F. Quintana-Ascencio. 2021. Integral Projection Models. R. Salguero-Gómez & M. Gamelon, (eds.). Demographic Methods across the Tree of Life, Oxford University Press. Oxford, Great Britain.

Goodman, L.A. 1969. The analysis of population growth when the birth and death rates depend upon several factors. Biometrics, 25, 659-681.

Gordon, D.R. and K. J. Rice. 1998. Patterns of differentiation in Wiregrass (Aristida beyrichiana): implications for restoration efforts. Restoration Ecology 6: 166-174.

Gross, C. 2016. Scientific misconduct. Annual Review of Psychology 67: 693-711.

Hanson, T. 2015. The triumph of seeds. Perseus Book Group. New York. New York.

Hilborn, R. and M. Mangel. 1997. The Ecological Detective. Princeton University Press. Princeton. New Jersey.

Hobbs, N.T. and M.B. Hooten. 2015. Bayesian Models: a statistical premier for ecologists. Princeton University Press. Princeton. New Jersey.

Howson, C. and P. Urbach. 1989. Scientific reasoning: the Bayesian approach. Open Court, La Salle, Illinois.

Hurlbert, S.H. 1984. Pseudoreplication and the design of ecological field experiments. Ecological Monographs 54: 187-211.

Jenkins, D. G., and P. F Quintana-Ascencio. 2020. A solution to minimum sample size for regressions. PLOS, 229345.

Johnson, J.B. and K. S Omland. 2004. Model selection in ecology and evolution. Trends in Ecology and Evolution 19: 101-108.

Jones, O.R., A. Scheuerlein, R. Salguero-Gómez, C. Giovanni Camarda, R. Schaible, B.B. Casper, J.P. Dahlgren, J. Ehrlén, M. Begoña García, E.S. Menges, P.F. Quintana-Ascencio, H. Caswell, A. Baudisch, J.W. Vaupel. 2013. Varieties of ageing across the tree of life. Nature, 505, 169-173 Doi:10.1038/nature12789.

Judson, O. 2002. Dr. Tatiana's sex advice to all creation: the definitive guide to the evolutionary biology of sex. Henry Holt and Company, New York, New York.

Kéry, M and M. Schaub. 2012. Bayesian population analysis using WinBugs. A hierarchical perspective. Elsevier.

Koltz, A.M., Burkle, L.A., Pressler, Y., Dell, J.E., Vidal, M.C., Richards, L.A. and S.M. Murphy. 2018. Global change and the importance of fire for the ecology and evolution of insects. Current opinion in insect science 29: 110-116.

LaFollette, Marcel C. 2000. "The evolution of the "scientific misconduct" issue: An historical overview (44535C)." Proceedings of the Society for Experimental Biology and Medicine 224: 211-215.

Larson, B. 2011. Metaphors for environmental sustainability: redefining out relationship with nature. Yale University Press. Ann Arbor, Michigan.

Levin, S. A. 1992. The problem of pattern and scale in ecology: the Robert H. MacArthur award lecture. Ecology 73: 1943-1967.

MacMillan, I.C., and R. G. McGrath. 1997. Discovering new points of differentiation. Harvard business review, 75: 133-142.

Maor, E. 2011. *e*: The story of a number, Princeton University Press.

Martin, T.G., B. A. Wintle, J.R. Rhodes, P.M. Kuhnert, S.A. Field, S.J. Low-Choy, A.J. Tyre, and H. P. Possingham. 2005. Zero tolerance ecology: improvement ecological inference by modeling the source of zero observation. Ecology Letters 8: 1235-11246.

Matthiopoulos, J., 2011. How to be a quantitative ecologist: the'A to R'of green mathematics and statistics. John Wiley & Sons.

McElreath, R.M. 2016. Statistical Rethinking: a Bayesian course with examples in R and Stan. Chapman and Hall.

McCarthy. M.A. and P. Masters. 2005. Profiting from prior information in Bayesian analysis of ecological data. Journal of Applied Ecology 42: 1012-1019.

McCarthy. M.A. 2007. Bayesian Methods for Ecology. Cambridge University Press.

Metcalf C.J.E., McMahon S.M., Salguero-Gómez R. and E. Jongejans. 2013. IPMpack: a R package for Integral Projection Models. Methods in Ecology and Evolution 4,195-200.

Merow, C., Dahlgren, J. P., Metcalf, C. J. E., Childs, D. Z., Evans, M. E., Jongejans, E., ... and S.M. McMahon. 2014. Advancing population ecology with integral projection models: a practical guide. Methods in Ecology and Evolution, 5, 99-110.

Menges, E. S. 1999. Ecology and conservation of Florida scrub. In R. C. Anderson, J. S. Fralish, and J. M. Baskin (Eds.), The savanna, barren, and rock outcrop communities of North America (pp 7–22). New York: Cambridge University Press.

Menges, E.S. 2000. Population viability analyses in plants: Challenges and opportunities. Trends in Ecology and Evolution, 15, 51–56. https:// doi.org/10.1016/S0169-5347(99)01763-2

Menges, E.S., Crate, S.J.H., and P.F. Quintana-Ascencio. 2017. Dynamics of gaps, vegetation, and plant species with and with- out fire. American Journal of Botany, 104, 1825–1836. https://doi. org/10.3732/ajb.1700175

Menges, E.S., and N.M. Kohfeldt. 1995. Life history strategies of Florida scrub plants in relation to fire. Bulletin of the Torrey Bot Club, 122, 282–297. https://doi.org/10.2307/2996320

Menges, E.S., Main, K.N., Pickert, R.L., and K. Ewing. 2017b. Evaluation of a fire management plan for fire regime goals in a Florida landscape. Natural Areas Journal, 37, 212–227.

Menges, E.S., and P.F. Quintana-Ascencio. 2004. Population viability with fire in *Eryngium cuneifolium*: Deciphering a decade of demographic data. Ecological Monographs, 74, 79–99. https://doi. org/10.1890/03-4029.

Miller, T., Quintana-Ascencio, P.F., Maliakal-Witt, S., and E.S. Menges. 2012. Metacommunity dynamics over 16 years in a pyro- genic shrubland. Conservation Biology, 26, 357–366. https://doi. org/10.1111/j.1523-1739.2011.01807.x

Nuzzo, R. 2014. Statistical errors. *P* values, the "golden standard" of statistical validity, are not as reliable as many scientists assume. Nature 506:150-152

Oakley, C.G. 2015. The influence of natural variation in population size on ecological and quantitative genetics of the endangered endemic plant *Hypericum cumulicola*. International Journal of Plant Science, 176: 11-19.

Oakley, C. G. 2013. Small effective size limits performance in a novel environment. Evolutionary Applications 6: 823-831.

Pearl J. and D. Mackenzie. 2018. The book of Why. The new Science of cause and effect. Basic Books. Hachette Book Company. New York, NY.

Quintana-Ascencio, P.F., Dolan, R.W., and E.S. Menges. 1998. *Hypericum cumulicola* demography in unoccupied and occupied Florida scrub patches with different time-since-fire. Journal of Ecology, 86: 640–651.

Quintana-Ascencio, P.F., Koontz, S.M., Smith, S.A., Sclater, V.L., David, A.S., and E.S. Menges. 2018. Predicting landscape-level distribution and abundance: Integrating demography, fire, elevation, and land- scape habitat configuration. Journal of Ecology, 106: 2395–2408.

Quintana-Ascencio, P.F. Koontz, S.M., Ochocki, B., Sclater, V. L., López-Borghesi, F., Li, H. and E. S Menges. 2019. Assessing the roles of seed bank, seed dispersal and historical disturbances for metapopulation persistence of a pyrogenic herb. Journal of Ecology, 107: 2760-2771.

Quintana-Ascencio, P.F., and E.S. Menges. 1996. Inferring metapopulation dynamics from patch-level incidence of Florida scrub plants. Conservation Biology, 10: 1210–1219.

Quintana-Ascencio, P.F. and E.S Menges. 2000. Competitive abilities of three narrowly endemic plant species in experimental neighbor- hoods along a fire gradient. American Journal of Botany, 87: 690–699.

Quintana-Ascencio, P.F., Menges, E.S., and C.W. Weekley. 2003. A fire-explicit population viability analysis of *Hypericum cumulicola* in Florida rosemary scrub. Conservation Biology, 17: 433–449.

Quintana-Ascencio, P.F., E. S Menges, G. S. Cook, J. Ehrlén, and M E Afkhami. 2021. Drivers of demography: past challenges and a promise for a changed future. R. Salguero-Gómez & M. Gamelon, (eds.). Demographic Methods across the Tree of Life, Oxford University Press. Oxford, Great Britain.

Quintana-Ascencio, P.F., and M. Morales Hernández 1997. Fire-mediated effects of shrubs, lichens and herbs on the demography of *Hypericum cumulicola* in patchy Florida scrub. Oecologia, 112: 267–271.

Quintana-Ascencio, P.F., Weekley, C.W., and E.S. Menges. 2007. Comparative demography of a rare species in Florida scrub and road habitats. Biological Conservation, 137: 263–270.

R Core Team. 2019. R: A language and environment for statistical computing. R Foundation for Statistical Computing, Vienna, Austria. URL https://www.R-project.org/.

Salguero-Gómez, R., Jones, O. R., Archer, C. R., Buckley, Y. M., CheCastaldo, J., Caswell, H., … and H. Buhr. 2015. The COMPADRE plant matrix database: An open online repository for plant demography. Journal of Ecology, 103, 202–218. https://doi.org/10.1111/1365-2745.12334

Shingleton, A. 2010. Allometry: The Study of Biological Scaling. Nature Education Knowledge 3: 2.

Smith, A.F.M. 1973. A general Bayesian linear model. Journal of the Royal Statistical Society: Series B 35: 67-75.

Stan Development Team. 2019. Stan Modeling Language Users Guide and Reference Manual, VERSION 2.26. https://mc-stan.org.

Stearns, S.C. 1999. The evolution of life histories. Oxford University Press. New York, New York.

Stengers, I. 2018. Another science is possible: a manifesto for slow science. Wiley.

Sujit. K.G., P. Mukhopadhyay and J.-C. Lu. 2006. Bayesian analysis of zero-inflated regression models. Journal of Statistical Planning and Inference 136: 1360-1375.

Touchon, J.C. and M.W McCoy. 2016. The mismatch between current statistical practice and doctoral training in ecology. Ecosphere 7: e01394.10.1002/ecs2.1394

Tye M R., Menges E.S, Weekley C., Quintana-Ascencio, P.F. and R. Salguero-Gomez. 2016. A demographic menage a trois: Interactions between disturbances both amplify and dampen population dynamics of an endemic plant. Journal of Ecology, 104:1778-1788

Vehtari, A., Gelman, A., and J. Gabry. 2017. Practical Bayesian model evaluation using leave-one-out cross-validation and WAIC. Statistics and Computing. 27: 1413–1432. doi:10.1007/s11222-016-9696-4. Online, arXiv preprint arXiv:1507.04544.

Wallace-Wells, D. 2019. The unhabitable earth: life after warming. Tim Duggan Books. New York.

Wasserstein, R. L. and N.A. Lazar. 2016. The ASA's statement on p-values: context, process and purpose. The American Statistician, Doi: 10.1080/00031305.2016.1154108.

Warton, D.I. and F.K.C. Hui. 2011. The arcsine is asinine: the analysis of proportions in ecology. Ecology 92: 3-10.

Watanabe, S., 2013. WAIC and WBIC are information criteria for singular statistical model evaluation. In: Proceedings of the Workshop on Information Theoretic Methods in Science and Engineering (pp. 90-94).

Zuur, A.F., E.N. Ieno, N.J. Walker, A. Savaliev, and G.M. Smith. 2009. Mixed effects models and extensions in Ecology with R. Springer.

Zuur, A, J.M. Hilbe and E N. Leno. 2015. A beginner's guide to GLM and GLMM with R. Highland Statistics, Ltd.

INDEX